QUANTUM COMPUTATION AND QUANTUM COMMUNICATION:
Theory and Experiments

QUANTUM COMPUTATION AND QUANTUM COMMUNICATION: Theory and Experiments

Mladen Pavičić
University of Zagreb
Zagreb, Croatia

Mladen Pavičić
University of Zagreb, Croatia

Consulting Editor: D. R. Vij

Library of Congress Cataloging-in-Publication Data

Pavičić, Mladen.
Quantum computation and quantum communication : theory and experiments /
Mladen Pavičić.
p. cm.
Includes bibliographical references.

ISBN 10 1-4614-9892-9
ISBN 13 978-1-4614-9892-6
ISBN 0-387-28900-3 (eBook)
1. Quantum computers 2. Quantum theory. I. Title

QA76.889.P38 2005
004.1—dc22 2005051725

Softcover re-print of the Hardcover 1st edition 2006

9 8 7 6 5 4 3 2 1 SPIN 11051145

springeronline.com

Dedicated to the reader

Contents

Preface

The attraction of quantum computation and quantum communication theory and experiments lies in the fact that we engineer both them themselves and the quantum systems they treat. This approach has turned out to be very resilient. Driven by the final goal of calculating exponentially faster and communicating infinitely more securely than we do today, as soon as we encounter a limitation in either a theory or experiment, a new idea around the no-go emerges. As soon as the decoherence "demon" threatened the first computation models, quantum error correction theory was formulated and applied not only to computation theory but also to communication theory to make it unconditionally secure. As soon as liquid-state nuclear magnetic resonance experiments started to approach their limits, solid-based nuclear spin experiments—the Kane computer—came in. As soon as it was proved that it is theoretically impossible to completely distinguish photon Bell states, three new approaches appeared: hyperentanglement, the use of continuous variables, and the Knill–Laflamme–Milburn proposal. There are many more such examples.

What facilitated all these breakthroughs is the fact that at the present stage of development of quantum computation and communication, we deal with elementary quantum systems consisting of several two-level systems. The complexity of handling and controlling such simple systems in a laboratory has turned out to be tremendous, but the basic physical models we follow and calculate for the systems themselves are not equally intricate. We could say that the theory of the field leads the experiments in a particular way—with each new model we put forward and apply in the laboratory, we also build up and widen the theory itself. Therefore, we cannot just proceed with assembling quantum computers and quantum networks. We also have to use mathematical models to understand the physics of each step on the road to our goal.

As a consequence, both mathematics and physics are equally essential for any approach in the field and therefore for this book as well. The mathematics used in the book is a tool, but an indispensable tool because the physics of quantum computation and communication theory and their experiments cannot be grasped without good mathematical models. When we describe an experiment many times, we may get used to it, but this does not mean we are more at home with the principles and models behind it. This is why I have chosen to make this book an interplay between mathematics and physics. The idea of the book is to present those details that are used the most often both in theory and experiment and to dispense with many inessential ones. Also, the book is not conceived as a textbook, at least not as a primary one, but more as a guide to a better understanding of theory and experiments by coming back to the same concepts in different models and elaborations. Clear physical ideas make any formalism easy.

MLADEN PAVIČIĆ

[ˈmlʌden ˈpʌvɪtʃɪtʃ][1]

[1]IPA, The International Phonetic Alphabet

Acknowledgments

There are a number of colleagues with whom I have discussed many details presented in this book and whom I have already thanked in person. There is one colleague, though, Norman D. Megill, who I would like to thank publicly for reading the manuscript repeatedly, making many substantive suggestions, and helping me to make it better. I am grateful for his expert advice.

The support of Series Editor D. R. Vij and the publisher team has been indispensable. I would like to express my gratitude to them.

I would also like to acknowledge the support of the *Ministry of Science, Education, and Sport* of Croatia through the project *Quantum Theory of Information* which I am the head of.

Introduction

Two predictions are cited particularly often whenever one talks or writes about the history or future of computing. One of these is more and more wrong, and the other is less and less right, and they both teach us how to use theoretical opportunities to find new technologies.

The first prediction, a beloved opening of speeches and papers, was made by the head of the electromagnetic relay calculator at Harvard, Howard Aiken, in 1956: "If it should turn out that the basic logics of a machine designed for the numerical solution of differential equations coincide with the logics of a machine intended to make bills for a department store, I would regard this as the most amazing coincidence that I have ever encountered" [Anonymous, 1997].

The amazing "coincidence" did happen and happens more and more every day, tempting us to consider it a part of the history of computers that took its own unexpected course ("Only six electronic digital computers would be required to satisfy the computing needs of the entire United States," Howard Aiken said in 1947): a program and a machine, software and hardware, were interwoven at the beginning and then became more and more separated. At least it seems so when we look at the development of computer designs since Charles Babbage's 1840s Analytical Engine. A program on punched cards or tapes and a machine for which the specific cards were made look inseparable, in contrast to today's programs which we move throughout the World Wide Web and compile and execute on virtually any computer.

Yet Alan Mathison Turing (and also Alonzo Church, Stephen Cole Kleene, and Emil Post independently at the same time) had already proved in 1936 that the only possible course the history could have taken was the one it in fact took. Turing used what we now also cite often and call a *Turing machine* to prove that only the simplest calculus, such as a propositional algebra with a Boolean evaluation (true, false) and its main model a 0–1 Boolean algebra, is computable, i.e., effectively calculable [Turing, 1936; Turing, 1937]. He (and others) also proved that real numbers are not computable, that there exists no algorithm with the help of which we can decide for every arithmetical sentence in finitely many steps whether it is true or false, etc. In other words, from the very start we only had Boolean algebra at our disposal, and once hardware was developed that could handle classical logic operations—such implementations of logic operations are called *logic gates*—the universal classical computer was born. The "only" thing one had to develop were "digital" algorithms and programs for all possible applications, i.e., the software for a universal computer. Everything—solving nonlinear differential equations, 3D modeling, speech recognition, and "making bills for a department store"—had to be reduced to a Boolean language. Since such a reduction imposes ever-growing speed and memory requirements upon the hardware, until mid-2002 we were witnessed quite the opposite situation than half a century ago: the software lagged behind the hardware, following the Wirth's law: "Software gets slower faster than hardware gets faster." Will this computing history repeat itself with quantum computers? Will quantum hardware start to advance faster than quantum software (quantum algorithms) in the near future? In this book we shall try to learn how close we are to answering these questions.

The second prediction is known as *Moore's Law*, or better yet, Moore's laws, since there are many versions and varieties of the several formulations made by Gordon Moore of the Intel Corporation. One widespread rendering of the law, "The number of transistors on a single integrated-circuit chip doubles every 18 months" [Birnbaum and Williams, 2000], does not correspond to the historical data which show 26 months [Brenner, 2001]. Moore himself commented. "I never said 18 months. I said one year [in 1965], and then two years [in 1975]. One of my Intel colleagues changed it from the complexity of the chips to the performance of computers and decided that not only did you get a benefit from the doubling every two years but we were able to increase the clock frequency, too, so computer performance was actually doubling every 18 months. I guess that's a corollary of Moore's Law. Moore's Law has been the name given to everything that changes exponentially in the industry... If Al Gore invented the Internet, I invented the exponential" [Yang, 2000].

And this "exponential" element is what is essential for our development and what quantum computers are about. Apparently everything underlying the development of technology and society grows exponentially: research, information, production and organization complexity, and above all, the costs of keeping pace. So only an exponential increase of our computational and processing power and an exponential decrease of computer cost per processed bit could support such a development. Therefore, Moore's law was been kept as a guideline in the computer industry in past three decades and it has supported a global development during this period.

Gates in today's computers are switched on and off by about 1000 electrons. In 2010, the exponential Moore's Law would require that only about 10 electrons do the job. Miniaturization cannot go much further than that. It is true that many other possible roads could still keep up the pace for a few more years: insulating layers can be reduced in their thickness from the present 25 atoms to 4 or 5 atoms (wires connecting transistors in a chip already occupy more than 25% of its space); computing power can be increased by designing processors so as to contain execution units that process multiple instructions within one cycle; processors can rely on parallel compiling technology and use innovative software; and finally, chips can eventually get bigger by using reversible gates to avoid overheating. Still, by 2020 or 2025 computing technology will hit the quantum barrier, and if we want to support the growth of our technology and science beyond that point in time, we need to find a substitute for exponentially rising classical computational power by then. Actually, the exponential increase of the clock speed of processors (CPUs) already became linear in 2002 (see Fig. 3.1, p. 135), and an extensive patching activity onto classical hardware and software is currently under way in order to compensate for this lack of an exponential increase in speed (see p. 136).

Now that both Wirth's and Moore's laws are coming to an end, we should draw a moral from them. Wirth's law taught us that classical hardware development has prompted ever new software, and Moore's law taught us that this hardware development has followed an exponential trend of speed, memory, and lately of number of processors (multiple cores, multiple processors, clusters). Such an approach to computation will apparently change completely in the quantum realm. Quantum hardware is exponential in itself, and if we eventually succeed in making functional scalable quantum computers, we will dispense with the need for a steadily growing quantum hardware development—to make a quantum computer faster means to scale it up linearly or polynomially. We will also dispense with writing ever new software for faster and faster hardware. Once developed, quantum software (quantum algorithms) will simply scale up as we scale—and therefore speed up—quantum hardware.

The "exponential" is built into quantum hardware from its very first *quantum bit* or *qubit*. Qubits, physically supported by single atoms, electrons, or photons, can superpose and entangle themselves so as to support an arbitrary number of states per unit. Recently devised algorithms—quantum software—relying on the exponential feature of quantum hardware have explicitly demonstrated how one can reduce important problems that are assumed to be exponentially complex, to polynomially complex tasks for quantum computers. This has opened a vast new interdisciplinary field of quantum computation and communication theories, together called quantum information theory, which along with its experimental verifications are already taught at many universities and have resulted in several very successful textbooks.

The target of these courses, seminars, and textbooks is to teach and familiarize students and scientists with this new field—in which new research projects will keep opening for decades to come—and to help integrate the theory and experiments of quantum computation and communication into a would-be quantum network implementation. The goal of the book in front of the reader is the same; however, it allows her or him to digest the field "by reading." That means that there will be no homework and no exercises. Instead, most of the required details are elaborated within the main body of the book, and a polynomial complexity of reading is intended, optimally in one run.

So, a few words about the reader. She or he is expected to be familiar with higher mathematics and the basics of physics—in particular, quantum physics. The reader could be any former student who graduated in the technical or natural sciences, although an undergraduate student might also find many if not all sections of the book digestible. Students as well as specialists in the field might also find the nutshell approach of the book helpful and stimulating.

Chapter 1

BITS AND QUBITS: THEORY AND ITS IMPLEMENTATION

In 1936 several authors showed, in effect, that if a function is effectively calculable, then it is Turing computable and, of course, vice versa [Church, 1936c; Turing, 1936; Turing, 1937; Church, 1936a; Church, 1936b; Kleene, 1936; Post, 1936]. Turing concluded:

> We do not need to have an infinity of different machines doing different jobs. A single one will suffice. The engineering problem of producing various machines for various jobs is replaced by the office work of "programming" the universal machine to do these jobs [Turing, 1948].

This statement does not mean that Turing envisioned the "universal computer" we have today, although he was well acquainted with the project of breaking the cryptographic codes of German messages carried out on the Colossus (the British "computer" at Bletchley Park, which operated from 1943 until the 1950s). His *universal Turing machine* is a "universal computer" only in the sense that it keeps to the standard digital (classical, 0–1) implementation, i.e., to the *bi*nary digi*ts*, or *bits*, of today's hardware.

1.1 The Turing Machine vs. a Computing Machine

The software used by any classical computer must be based on what a Turing machine can confirm to be calculable, recursive, and decidable. A historical problem with the development of computers was that there were few calculus categories of the latter kind. The only types of calculus that Turing machines can show to be calculable are the simplest algebras with the simplest evaluations, such as propositional calculus with Boolean (true–false) evaluation, or 0–1 Boolean algebra. It can be shown that even the simplest propositional calculus with a nonordered evaluation[1] [Pavičić and Megill, 1999] or simplest arithmetic with natural numbers [Hermes, 1969] is not calculable simply because such types of algebra are neither recursive nor decidable nor calculable. Directly, a Turing machine can only be used to *prove* that no mathematics we know from primary school can be literally run on it.

Turing machines, or any equivalent mathematical algorithms, are essential in order to decide whether a chosen problem is calculable or not, but we do not use them to write down a new program for, say, 3D modeling or speech recognition. Still, since there are many references to the Turing machine in the literature on quantum computing, let us provide some details [Hermes, 1969]. In doing so, we bear in mind that Turing machines and all related concepts are "concepts of pure mathematics. It is however very suggestive to choose a technico-physical terminology suggested by the mental image of a machine" [Hermes, 1969, p. 31].

The Turing machine is neither today's "universal" computing machine—generally called a computer—nor a generator of new algorithms for the latter machine. Instead, it is simply a mathematical procedure to check whether a chosen algebra and/or calculus can or cannot be implemented into a computer. To show this, we present some details of the procedure. The details often appear in the literature without being put into the context of a final outcome and so are just left hanging, giving the impression of being building blocks for a computer, or an algorithm to be carried out on one. On the other hand, the notion of the classical Turing machine is rather important for understanding the role that the quantum Turing machine has in the theory of quantum computation.

1.2 Definition of a Turing Machine

We start with an alphabet $\mathcal{U} = \{a_1, \ldots, a_N\}$, $N \geq 1$. There is also a blank symbol a_0 that does not belong to the alphabet. We then define a *Turing machine* M *over* $\mathcal{U}$ as given in Table 1.1, where $c_1, \ldots, c_M$ are different natural numbers ≥ 0 and $\mathcal{S} = c'_j \in \{c_1, \ldots, c_M\}$ for $j = 1, \ldots, MN$. Furthermore, $b_j \in \{a_0, \ldots, a_N, r, l, h\}$, where r, l, and h do not belong to $\mathcal{U}$ and, as we will see later on, refer to *right*, *left*, and *halt*. c_j's are called *states*, and c_1 is called the initial state. $\{a_0, \ldots, a_N\}$ are values $B(x)$ of *function* B, where x is an integer.

[1] In such a calculus one can ascribe value α to one proposition and β to another in a consistent way so as to build a model for a Boolean algebra, while neither $\alpha < \beta$, nor $\alpha = \beta$, $\alpha > \beta$ holds.

To obtain a visualization of a machine, we interpret x's as numbers of *squares* of a *computing tape* arranged successively, as shown in Table 1.2. The square with the number n lies immediately *to the left* of square $n+1$.

It is due to this visualization that Table 1.1 has been named a *machine.* We obtain the result of an operation, say addition, as a number expressed by an encoding of the symbols $\{a_1, \dots, a_N\}$ positioned somewhere on the tape. Thus operations on numbers and the procedures for obtaining final outputs look as if a scanning device "moves" left and right over the tape (see Table 1.2).

Table 1.1. Turing machine. The four columns are the input state, the input value, the output "action," and the output state (where $c'_j \in \{c_1, \dots, c_M\}$ for $j = 1, \dots, MN$). Actions r, l, and h are called *left, right,* and *halt*; an action $a_i \in \mathcal{U}$ denotes "print the symbol a_i;" and action a_0 denotes "erase." State c_1 is called the *initial state.*

c_1	a_0	b_1	c'_1
...	...	...	...
c_1	a_N	b_{N+1}	c'_{N+1}
c_2	a_0	b_{N+2}	c'_{N+2}
...	...	...	...
c_2	a_N	b_{2N+2}	c'_{2N+2}
...	...	...	...
...	...	...	...
c_M	a_N	b_{MN+M}	c'_{MN+M}

We assume that square n has the symbol $B(n)$ *printed* in it; for example, square -2 has the symbol a_4. A square that has the symbol a_0 printed in it is called *empty* and is left blank in Table 1.2. Thus the function B is a *tape expression* of the computing tape. The tape can be infinite, but we assume that there are only a finite number of symbols a_i, $i = 1, 2, 3, \dots$, printed on it. Other squares are empty.

Table 1.2. Computing tape

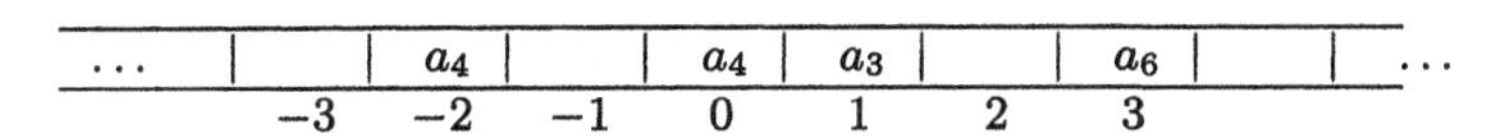

...		a_4		a_4	a_3		a_6		...
	-3	-2	-1	0	1	2	3		

A *configuration* K of a Turing machine M is defined as an ordered triple $K = (A, B, C)$, where A is a square (given by its number), B is a tape expression (a function), and C is a state of M. In the aforementioned "machine language," we say that we *place the machine* M *on the tape expression* B *over the square* A. C determines whether a configuration is initial, consecutive, or terminal. In the last case, we simply say that the machine stops or *halts*.

Referring to the symbols introduced at the beginning of the section and taking into account the above details, we can briefly restate the definition of a (deterministic) Turing machine as follows.

DEFINITION 1.1 *A* deterministic Turing machine *is a triple* $(\mathcal{U}, \mathcal{S}, \delta_{\mathrm{d}})$, *where* δ_{d}, *the* deterministic transition function, *is a function*

$$\delta_{\mathrm{d}} : \mathcal{U} \times \mathcal{S} \longrightarrow \mathcal{U} \times \mathcal{S} \times \{l, r, h\}. \tag{1.1}$$

To see how the machine works, let us first limit the alphabet $\mathcal{U}$ to only one symbol, *tally* "ı." Next, we limit our objective to calculation of numerical values of functions. In doing so we make use of the following correspondence: 0 is ı, 1 is ı ı, 2 is ı ı ı, etc. We, being *human computers* [Turing, 1950] (see footnote 2, p. 6), write down our result on an empty sheet of paper (tape).

To place the machine (alternative terminology: control unit, read-write unit, head, read–write head) over a square, read a tape expression in the square, and move to another square or stop means literally that if in the first of the c_1 lines in Table 1.1, in the second column there is the symbol a_0, i.e., an empty symbol, then we carry out the command b_1, and we go on to carry out the line c_1' and so on repeatedly. To carry out b_1 means that we write down—on the tape, which is our "output"—a tally, or we leave it blank (erasing it if necessary), or we move it to the left or right, or that we stop computing. An example of a Turing machine that produces an infinite number of zeros is given in Table 1.3 [Hermes, 1969, p. 40].

Table 1.3. Example of a Turing machine that prints an infinite sequence

0	a_0	ı	0
0	ı	r	1
1	a_0	r	0
1	ı	h	1

1.3 Turing Computability

In the example shown in Table 1.3 if we place a Turing machine over an arbitrary square of an empty tape, it will read the symbol in the square (a_0, blank symbol); carry out the command in the third column, i.e., print ı, and move to $c_1 = 0$, i.e., to the next row. In the next row, we repeat the procedure: the Turing machine will read the symbol in the square, ı, carry out the command in the third column, i.e., move one square to the right, and move to $c_2 = 1$, i.e., to the third row. In the third row, the Turing machine will read the symbol in the square, a_0; carry out the command in the third column, i.e., move one square to the right, and move to $c_1 = 0$, i.e., to the first row. And so on, ad infinitum: this Turing machine will never visit the fourth row, i.e., it will never stop. The outcome is an infinite sequence of zeros, since ı corresponds to a zero. However, Turing machines that do not stop, or more precisely Turing machines that do not stop on particular tapes fed to them, are not important and play no role in the theory of computation and computing. Only Turing machines that can calculate functions when acting on chosen tapes, i.e., which eventually stop, are of importance.

A function f is *Turing computable* if there exists a Turing machine M such that, if we print an arbitrary expression x onto an otherwise empty tape and if we place M over an arbitrary square of the tape, then M will stop operating after finitely many steps *behind* (over a sequence of making tallies) an expression that represents the value $f(W)$ of the function [Hermes, 1969]. For instance, if $f(3,0) = 1$ and M computes f, then when M is started in the configuration

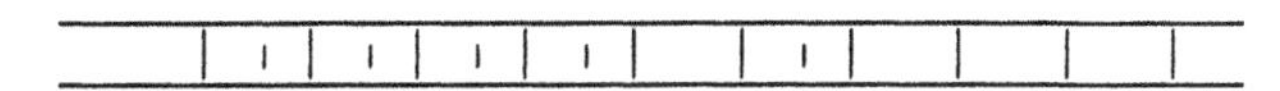

it will stop in the configuration over the last tally on the right-hand side, i.e., behind the last and the next-to-last tallies meaning number 1:

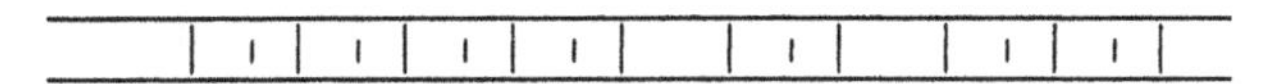

In other words, a number-theoretical function f is called *Turing computable* if there exists a Turing machine that can compute its value. Turing computability enables us to give every problem that can be formulated for Turing machines a mechanically obtainable solution according to a fixed routine (set of instructions).

Turing machines are not a unique way to express computability. The following formulations have been proved to be equivalent:

- *Turing computability* [Turing, 1936],
- *general recursiveness* [Herbrand, 1931; Gödel, 1931; Gödel, 1934],
- *λ-definability* [Church, 1936c],
- *μ-recursiveness* [Kleene, 1936],
- *reckonability* [Gödel, 1934],
- *evaluation according to rule* [Hilbert and Bernays, 1934; Hilbert and Bernays, 1939],

- *binormality* [Post, 1943],
- *normal algorithm* [Markov, 1947; Markov, 1961].

The conjecture that any of these definitions of computability is equivalent to effective calculability is known as *Church's thesis*. There is overwhelming evidence for the correctness of the conjecture. Together with the notions of computability and recursiveness, i.e., calculability, we also introduce the notion of *decidability*. A theory or a calculus or an algebra $\mathcal{A}$ is *decidable* if there exists an effective, uniform method of determining whether an expression formulated in $\mathcal{A}$ is valid in $\mathcal{A}$. Of course, a highly desirable property of any theory is that we are eventually able to "calculate" any expression of the theory and to decide on its decidability in an automated, mechanical way according to a fixed recipe. Turing's approach resembled this idea, and for students of formal logic and foundations of mathematics, it was the easiest to grasp. Therefore Turing machines are often used as a bridge between a calculus and a computer on which we can "calculate the calculus."[2]

Courses on the theory of computation often start with Turing machines and then jump to the architecture of a modern computer without a further reference to Turing machines. This is because the Turing machines are just the tip of a mathematical iceberg. Historically, Turing computability, general recursiveness, or any other of the aforementioned alternatives played a major role in solving the *decision problem* for *validity* and *provability* of formal theories, i.e., whether a given formula is valid and whether there is a proof for a theorem. The easiest way to understand this is by means of *Gödel numbers*. To any primitive symbol such as $=$, $\in$, $\forall$, x, etc., we can ascribe a Gödel number 1, 2, 3, 4, etc., respectively, and therefore can also do the same to a sequence of symbols that make formulas.[3] Thereby the notions of derivation, proof, etc. become representable by numbers. To see whether a series of formulas is a proof of a given formula, we use a Turing machine or any of its equivalents as a mechanical procedure of checking the formulas. In a word, "all syntactic questions such as whether a given formal system is decidable are mirrored into corresponding number-theoretical questions such as whether a certain number-theoretical property is general recursive [Turing computable, λ-definable, etc.]" [Fraenkel et al., 1973, p. 308].

[2] "The idea behind digital computers may be explained by saying that these machines are intended to carry out any operations which could be done by a human computer [a human being who calculates unaided by any machinery save paper and pencil]." [Turing, 1950]

[3] One can even carry out a Gödel numbering of Turing machines themselves [Hermes, 1969].

What makes a theory calculable on a computer is nothing but its aforementioned computability, i.e., its decidability. An apparent similarity between a computer routine and a Turing machine has little to do with a decidability proof of a theory. To carry out such a proof for even a simplest theory by means of Turing machines, we need to define a number of machines (left, right, search, copying, etc.) [Hermes, 1969], combine them together, then use a number of symbols and states (two or more), and in the end carry out proofs of quite a number of nontrivial theorems on hundreds of pages.[4] This does not mean, though, that we could not use Turing machines for some down-to-earth algorithms such as multiplying matrices. We could, but if we want to obtain the resulting matrix quickly, we had better use Boolean circuits, in plain words: a computer.

1.4 Bit Computability: Boolean Algebra

After trying to prove decidability and calculability for simple theories at the beginning of the twentieth century, mathematicians were surprised by how few proofs turned out to be possible. Even the simplest arithmetic and predicate calculus turned out to be undecidable and "noncalculable." Theories (systems, algebras) that had decidability proofs were Boolean algebra and a few with Boolean algebra as their model. An example of the latter category is the classical propositional logic, i.e., a propositional logic but with only its numerical evaluation (true–false, 0–1, n-valued). The propositional logic—as far as only its axioms are concerned—can, however, also have a nonnumerical evaluation, for which it cannot have a Boolean algebra as a model. The model for such an evaluation is an algebra that is not distributive and which therefore cannot be implemented into classical computer hardware [Pavičić and Megill, 1999].

Hence any classical computer is built so as to make use of Boolean algebra in the binary {0,1} evaluation. Essentially, the algebra can follow directly from a single universal operation known as the *Sheffer stroke* $(A|B)$ in the logic literature and as NAND in the computing literature, where A and B are the inputs and $A|B$ is the output.

[4] *Universal Turing machines*—those that can mimic any chosen Turing machine—do not contribute to such proofs. So, at least for the time being, the universal Turing machines are just puzzles in themselves. The smallest known such machines are of the following *complexities*: $\{24, 2\}$ and $\{2, 5\}$, where the first number is the number of states and the second the number of symbols. $\{2, 5\}$ has recently been found by [Wolfram, 2002, p. 707]. $\{2, 4\}$ and $\{2, 3\}$ have also been conjectured [Wolfram, 2002, p. 708,9], but $\{2, 3\}$ apparently contradicts Theorem 4.1.21 from [Gruska, 1997, p. 226] which sets the minimal complexity to "no smaller than 7" because $2 \cdot 3 = 6 < 7$.

If we use this single universal operation, there is even a way to define the Boolean algebra with the help of a single axiom. The NAND can therefore be used to express all the standard Boolean operations: NOT ($\overline{\ }$, $\neg$), AND ($\cdot$,)[5], OR (+), and XOR ($\oplus$) [XOR is addition (+) modulo 2: $1 \oplus 1 = 0$; also $A \oplus B = (A + B)\overline{AB}$]. For instance,

$$\overline{A} = A|A, \qquad AB = (A|B)|(A|B), \qquad A + B = (A|A)|(B|B). \tag{1.2}$$

NAND itself can be expressed with the help of NOT and AND, or NOT and OR as follows:

$$A|B = \overline{AB} = \overline{A} + \overline{B}. \tag{1.3}$$

The other such universal operation is NOR: $A \downarrow B = \overline{\overline{A}|\overline{B}}$. Only NAND and NOR can be used to define Boolean algebra with a single operation.

For practical implementation of Boolean algebra, though, short axiomatics are not particularly useful because we need simple axioms to implement shortcuts in calculations. One could ask: Do we need an algebra at all? Could we not carry out all the calculations by just using 0s and 1s and the operations defined on them, without bothering about any underlying algebra and axiomatics? In other words, could we not just use the so-called *truth tables* as shown in Table 1.4 to define our operations?

Table 1.4. NOT ($\overline{\ }$), NAND (Sheffer stroke), AND, OR, and XOR operations

A	B	$\overline{A}$	$A\|B$	AB	$A+B$	$A \oplus B$
0	0	1	1	0	0	0
0	1	1	1	0	1	1
1	0	0	1	0	1	1
1	1	0	0	1	1	0

Let us check this idea and at the same time see how a computer handles 0s and 1s. To verify an expression with 30 variables (even a simple 16-bit adder that sums up numbers written down in binary code has more than 30 variables), we need to check over 30 billion truth values. As opposed to this process, the following axioms [Langholz et al., 1989] give the answer in a few lines and within a fraction of a second.

DEFINITION 1.2 *A* Boolean algebra *is an algebraic structure consisting of a set of elements $\mathcal{B}$ together with two* binary *operations,* addition (+)

[5]We usually drop $\cdot$ and write, for example, AB instead of $A \cdot B$.

and multiplication ( ), *and a* unary *operation,* negation ($\overline{\ }$), *such that the following axioms hold (for every* $A, B, C \in \mathcal{B}$*):*

1. *The set* $\mathcal{B}$ *contains at least two elements* A, B *such that* $A \neq B$.
2. Closure property (a) $A + B \in \mathcal{B}$, (b) $AB \in \mathcal{B}$
3. Commutativity (a) $A + B = B + A$, (b) $AB = BA$
4. Existence of identities (0,1) (a) $A + 0 = A$, (b) $A1 = A$
5. Distributivity (a) $A + BC = (A + B)(A + C)$, (b) $A(B + C) = AB + AC$
6. Existence of complements (a) $A + \overline{A} = 1$, (a) $A\overline{A} = 0$

The aforementioned equivalent definition based on the NAND uses, for example, nothing but the axiom

$$(A|((B|A)|A))|(B|(C|A)) = B, \tag{1.4}$$

which has been proven to be the shortest among such axioms by William McCune [McCune et al., 2002]. A number of single axioms have recently been found by Stephen Wolfram [Wolfram, 2002, pp. 808,1174].

To find an algebra for quantum mechanics—analogous to the Boolean algebra for classical mechanics—to be run on quantum computers might prove to be a major challenge in the future. But let us go back and see how a classical computer handles 0s and 1s.

1.5 Bit Implementation: Transistors and Their Limits

Electrical 0–1 switches in today's computers are mostly metal-oxide semiconductor field effect transistors, or MOSFETs. They are made of silicon semiconductors with impurities called dopants. If silicon, which has four valence electrons, is doped with arsenic, which has five valence electrons, the resulting material will contain free electrons, and we say that the obtained material is negatively N-doped. If doped with boron, which has three valence electrons, the material will show a lack of electrons described by positive "holes," so we say the material is P-doped. Two types of MOSFETs are the NPN MOSFET called NMOS, shown in Fig. 1.1, and the PNP MOSFET called PMOS. An NMOS transistor consists of two N-doped regions (source and drain) of silicon that are slightly separated on a P-doped substrate. The substrate is called the *channel* because when the device turns "on," i.e., where the gate voltage is increased and the positively charged holes in the P substrate are repelled away from the insulator (oxide barrier) and electrons are attracted to it, the current starts to flow from the drain to the source (technical direction; opposite to the electron flow). Actually, the drain and the source are interchangeable. A PMOS transistor, on the other hand turns "off" when the gate voltage is increased.

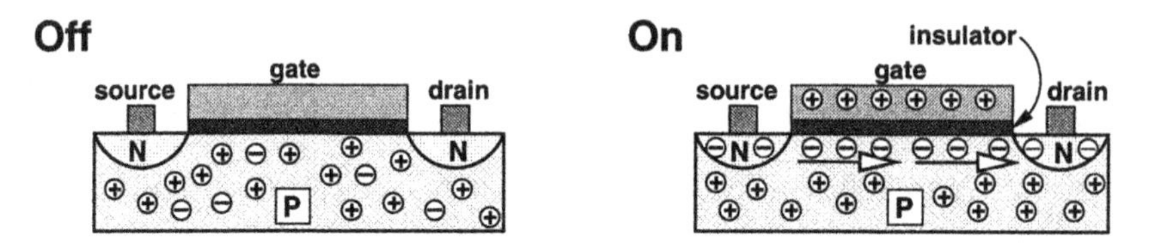

Figure 1.1. NPN MOSFET — NMOS (PMOS has N and P inverted)

By means of 0–1 switches, and logic gates derived from them, today's universal computer carries out all tasks, i.e., all operations such as *shift* (an operation that moves bits stored in a register to a new position); *move*; arithmetical and logical operations; conversions of representations of alphanumeric signs, real numbers, graphics, etc.; the management of output, input, and storage devices; etc. Early computers dissipated a lot of energy (heat) by using such transistor switches.

To see what caused dissipation and how it was reduced, let us consider the NMOS, PMOS, and CMOS circuits of Fig. 1.2. All three circuits are put together so as to build the NOT gate. In the circuits, the NMOS and PMOS transistors are connected to the positive supply and ground via pullup and pulldown resistors, respectively. CMOS dispenses with these resistors. To be able to compare the three circuits, we assume that they all have to operate between two voltage levels +V and the ground and that the information flows between In and Out of each gate. (Within real circuitry, these are parts of subcircuits; for example, not each such transistor is necessarily connected to the ground or +V.)

When a positive voltage (in 0–1 logic, In=1) is present at its gate, the NMOS transistor conducts electricity (is "on"). The current flows through the resistor, and the drain voltage is driven towards the ground. The output voltage is therefore zero volts (logic: Out=0). We say that the transistor is *closed* whenever current flows through it—like a switch. When the NMOS transistor is "off" (no voltage at its gate), no current flows through it, so its drain voltage and Out is positive (logic 1: Out=1). We say that the transistor is *open* when no current flows through it—like a switch.

In the PMOS circuit of Fig. 1.2, where the source is connected to the ground through a pulldown resistor, the reverse is true. When there

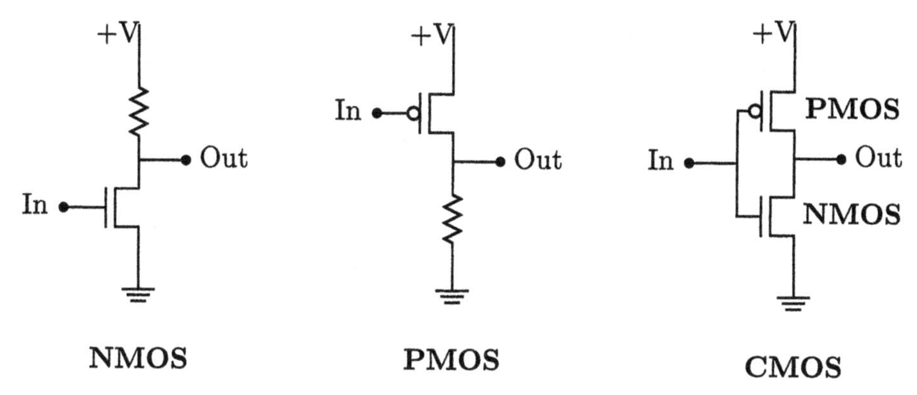

Figure 1.2. NOT gates. Older PMOS and NMOS devices dissipate heat through resistors, while the CMOS gate dispense with resistors. In all three cases, when In=1, then Out=0, and when In=0, then Out=1.

is no voltage at its gate, In=0, the PMOS transistor conducts and the output voltage is driven to +V: Out=1. The PMOS transistor is closed. A small current flows through the resistor to the ground. The power dissipated by the resistor is wasted, and this is why it is important for the resistance to be as high as possible. On the other hand, if the resistance is too high, the discharge of the parasitic capacitances associated with the transistor and surrounding circuitry it will take a long time, and the maximum operating speed (switching rate) of the circuit will be limited. When there is a positive voltage on its gate, the PMOS transistor does not conduct, and the resistor pulls the output voltage to the ground. The PMOS transistor is open.

The CMOS circuit of Fig. 1.2 combines NMOS and PMOS circuit but dispenses with the resistors. When a positive voltage is present at the input (In=1), the PMOS transistor turns off and the NMOS turns on, providing an insulating path to +V and a conducting path to the ground, so that the output voltage is driven to ground (Out=0). When the input voltage is zero (In=0), the PMOS transistor turns on and the NMOS turns off, providing a conducting path to +V and an insulating path to the ground so that the output voltage is driven to +V (Out=1). We do not need resistors because whenever one transistor is off the other is always on, providing a conducting path to either the ground or +V to rapidly (dis)charge parasitic capacitances for high-speed operation. But once the (dis)charging is completed, the CMOS circuit draws no current at all (unlike the circuits with resistors) until the next input voltage transition occurs (say, at the beginning of the next CPU clock cycle). This technological solution dispenses with great deal of the energy dissipation in transistor circuits, which, at the time when Richard Feynman [Feynman, 1985] started to think of a quantum computer, was one of the main worries with classical computers.

Other thermodynamic problems (e.g., energy dissipation) with classical computers might be solved eventually. For example, today's classical computers require hundreds of electrons per gate so as to be reliable. Apparently this number can be reduced to a hundred by using the present type of transistors. There are already substitutes for CMOS such as the *single electron transistor*, SET, based on tunnel junctions [Devoret et al., 1992] and quantum dots that can bring the number of electrons down to a few per gate. There is also a problem with the conductors that connect transistors. In today's processors, the conductors already occupy 30% of available space. With fewer electrons per gate, the conductors can be made thinner, but only down to monolayers (one-molecule-thick layers of conductors and insulators). Once these limits are reached, further shrinking will not be possible. One can still improve the performance of a classical computer with a few electrons, or even photons (although optical computers are still under development), per bit by increasing the size of the processor. But this approach is limited for two reasons: first, the number of elements necessary for a linear increase in performance grows exponentially; and, second, we will still have a thermodynamic minimal energy wasted per calculated bit that increases with shrinking. Let us look at the details.

1.6 Irreversible Bits: Logic Gates

Logic gates are the elementary building blocks of a digital transistor circuit and correspond to Boolean operations. As we have already mentioned, all Boolean operations can be expressed by means of NAND ($|$), but the expressions become simpler if we are allowed NOT ($\overline{\ }$ or $\neg$) ($\overline{\mathbf{A}} = \mathbf{A}|\mathbf{A}$ as well. For instance, the AND operation ($\cdot$) then reads $\mathbf{C}=\mathbf{A}\cdot\mathbf{B}=\mathbf{AB}=\overline{\mathbf{A}|\mathbf{B}}$, and OR ($+$) is $\mathbf{C}=\mathbf{A}+\mathbf{B}=\overline{\mathbf{A}}|\overline{\mathbf{B}}$. CMOS AND is then CMOS NAND followed by CMOS NOT. There is one important difference between the unary NOT operation and the binary operations (NAND, AND, etc.), though. We can represent $\mathbf{A}$ and $\mathbf{B}$ by column matrices whose entries are the two possible 0–1 values and the operation NOT in a 2×2 matrix, as in Eq. (1.5):

$$\mathbf{B} = \begin{bmatrix} 1 \\ 0 \end{bmatrix} = \begin{bmatrix} 0 & 1 \\ 1 & 0 \end{bmatrix} \begin{bmatrix} 0 \\ 1 \end{bmatrix} = \neg \begin{bmatrix} 0 \\ 1 \end{bmatrix} = \overline{\mathbf{A}} \tag{1.5}$$

We also have $\mathbf{A} = \overline{\mathbf{B}} = \neg\mathbf{B}$, which means that the NOT operation is reversible. On the other hand, no binary Boolean operation is reversible. There is no operation that can take us from $\mathbf{C} = \mathbf{AB}$ back to $\mathbf{A}$ and $\mathbf{B}$.

For example, if **C**=1—i.e., a nonzero voltage is detected at the output **C**, as in Fig. 1.3—we cannot know whether it was obtained by **A**=1 and **B**=0 or by **A**=0 and **B**=1 or by **A**=0 and **B**=0.

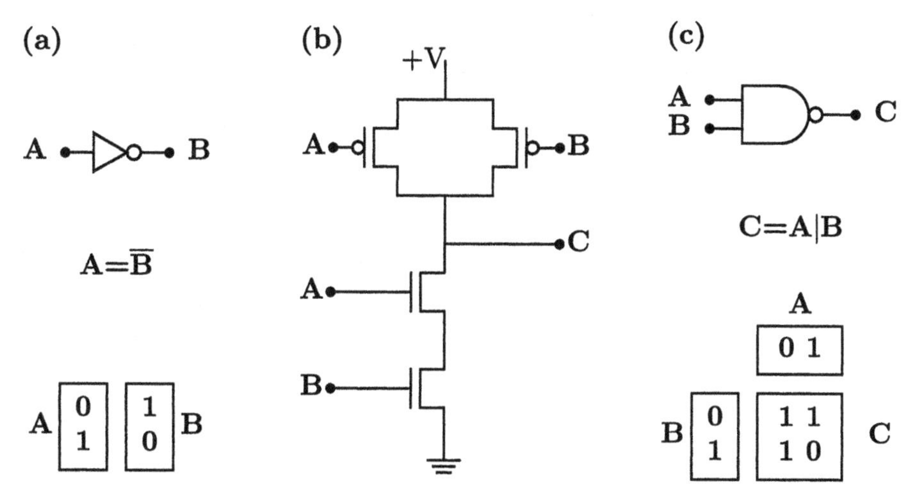

Figure 1.3. (a) NOT gate symbol, operation, and truth values; (b) CMOS NAND gate; (c) NAND gate symbol (note the "negating" bubble—the AND gate symbol lacks it), operation, and truth values.

The literature usually claims that this irreversibility is responsible for the lost energy per calculated bit. While it is certainly true that "logical computations that are not reversible necessarily generate heat, i.e., $kT \ln 2$, where k is Boltzmann's constant [1.38×10^{-23} J/K] and T the temperature in Kelvins, for every bit of information that is lost" [De Vos et al., 2002], in today's CMOS processors both reversible and irreversible logical computations generate the same amount of heat for every bit (per electron that carries the bit: one bit in today's computers is carried by up to 1000 electrons; the heat also depends on the voltage through which the electrons move). One can see this process in Figs. 1.1 and 1.2 (p. 10): in the present computer architecture, the currents that switch the gates are simply dumped to the ground or to +V, regardless of whether the operation is reversible or not. The reversibility of logical operations is only a necessary condition for a possible reuse of these currents with the help of a completely different would-be computer architecture. There have been several hardware proposals recently, not only ideal systems, such as billiard balls and Brownian-motion systems but also CMOS prototypes [De Vos et al., 2001]. A promising way to implement reversed directions of computation is by resonating the power rails, a process that simultaneously replaces the clock. This or some other design reduces to a scheme of a synchronized clock (resonator) pulse, which redirects currents but always according to an algebra of reversible computing. For present designs, the slower the clock, the more efficient the computing—and what we would like to have is ever-faster processors. Hopefully, more efficient designs will be found in the next few years, since by then reversible computing apparently will have to be implemented for the reasons given below.

1.7 Reversible Gates

When we calculate $kT \ln 2$ for, say, 100° C, we obtain 3.6×10^{-21} J(oules). This amount may seem negligible, but even with the absolute minimum of one bit of logical information supported by only one bit of physical information (electron state) in a gate, such a bit passes through thousands of gates during each of billions of clock cycles each second. When we consider that the number of transistors in a CPU (central processing unit) may soon reach 1 billion, and that within a few years CPU clock speed may exceed 10 GHz (see p. 135), we can estimate that the maximal number of processed bits per second will exceed 10^{19} bits per second per CPU. Then we can easily calculate that unless we overcome $kT \ln 2$ per bit, no cooling could prevent such CPUs from melting. As we have already said, the hardware will have to be changed to replace today's clock pulses with resonating "swings." For example, electrons could ballistically oscillate in silicon crystals or carbon nanotubes until they hit a programmed lattice defect, which change change the paths of some of them in order to perform logical operations. Such oscillations would need only a tiny amount of energy dissipation to keep them swinging with a constant frequency.

Reversible hardware will also require a new kind of software to implement and use a reversible logic algebra. It seems that development of such a software is feasible. For example, it has already turned out that a comparatively small number of oscillation delays would be required. In today's computers, a series of delays of relevant clock pulses is always implemented. In Figs. 1.1 and 1.2 we can see that a pulse (square wave) cannot arrive at both a gate and a source at the same instant. We first have to switch the gate on, and only after a delay can we let current through. Each transistor has such a delay built into it. Groups of gates in a computer are incorporated into sophisticated timed circuits. The gates then "calculate" within the time window determined by successive clock ticks. Any voltage-level change that occurs in response to a clock tick must charge or discharge parasitic capacitances associated with a transistor and its surrounding circuitry. The energy cost of (dis)charging a capacitor is $CV^2/2$. In a conventional circuit, most of this energy is dissipated resistively into the environment. In a reversible gate, on the other hand, the energy stored in parasitic capacitances is not dissipated but is returned back to the circuit.

The software for the first reversible processors has already been implemented [De Vos et al., 2001]. Essentially, it is a clever way to implement calculations by swinging electrons back and forth. At the end of each swing, all obtained outputs are either copied and taken out [Bennett, 1973; Bennett, 1989] or reintroduced into the next swing. A classical reversible computer will most probably be a link between today's classical computer and a would-be quantum one, when the shrinking of computer elements hits the one atom barrier in about two decades. It is no coincidence that reversible algebras underlying reversible computer and quantum computer theories were developed in parallel and that they have many characteristics in common: reversibility, control, and universality. General software, on the other hand, diverged: classical reversible computing software is just a technical blueprint for speeding up already existing general purpose hardware, while a general software for quantum computing still does not exist and is not likely to resemble reversible computing software. Therefore, we will not consider reversible computing software any longer but will present some details of its algebra that will turn out to be relevant for the algebra of quantum gates later on.

We mentioned above that the logically reversible NOT gate is not reversible physically (in today's computers) because a voltage must switch the gate in order for a current to pass through the source to the drain. So it might be better to call the gate states within reversible computers *before* and *after* rather than *input* and *output* as with standard computers. These terms stress that we do not let gate current "through" a transistor—we only redirect it so as to be able to reuse it. (This is why reversible computing is sometimes called *green computing.*)

In Fig. 1.4, we show the so-called controlled-NOT, CNOT operation, which reuses the "gate current" of a NOT gate and serves as a *reversible logic gate.* CNOT cannot be used to express and then reverse the NAND operation (see Fig. 1.3, p. 13 and Fig. 1.4 (a)) since it has only two outputs and for a reversible NAND we need three outputs: one to give its value and two to record the two inputs. CCNOT, shown in Fig. 1.4(b), can do just that—as shown in Fig. 1.5 (a) [Feynman, 1985]. Moreover, it turns out that CCNOT is one of over 38,975 universal logic gates [De Vos et al., 2001] among 8!=40,320 reversible gates with three binary inputs and three binary outputs.

Graphical representations of the kind presented in Fig. 1.4—reversible circuits—are very common in both reversible and quantum logic, where "logic" simply means a set of rules for handling gates. Expressions

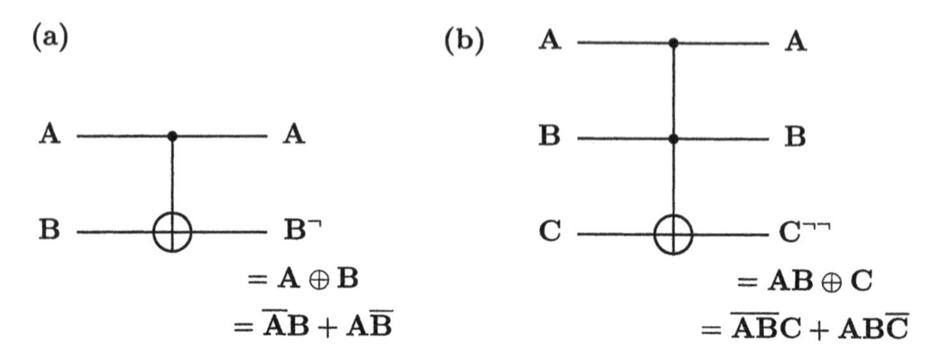

Figure 1.4. Reversible circuits: (a) C(ontrolled)NOT gate ($\mathbf{B}^{\neg} = \overline{\mathbf{B}}$ for $\mathbf{A} = \mathbf{1}$ and $\mathbf{B}^{\neg} = \mathbf{B}$ for $\mathbf{A} = \mathbf{0}$); (b) CCNOT (Toffoli) gate ($\mathbf{C}^{\neg\neg} = \overline{\mathbf{C}}$ for $\mathbf{A} = \mathbf{B} = \mathbf{1}$; $\mathbf{C}^{\neg\neg} = \mathbf{C}$ otherwise).

on the left-hand side are "before" (inputs) and on the right-hand side are "after" (outputs). Dots stand for controlling gates and mean that expressions do not change by passing through these gates from left to right. The gate they control is the $\oplus$ gate. Graphs can be concatenated as shown in Fig. 1.5 (b).

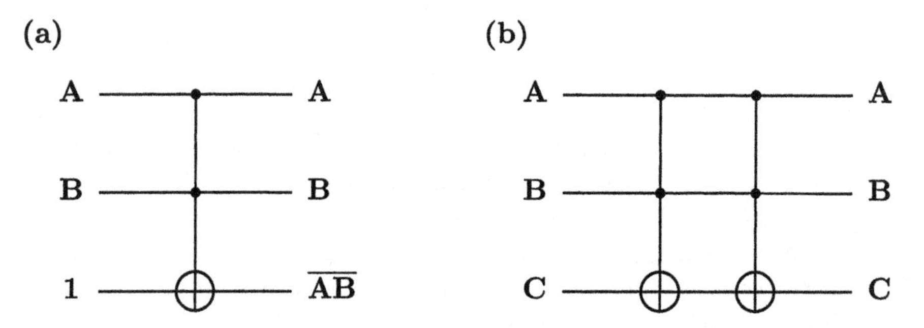

Figure 1.5. (a) NAND gate expressed by means of CCNOT gate: $\overline{\mathbf{AB}}\mathbf{1} + \mathbf{AB}\overline{\mathbf{1}} = \overline{\mathbf{AB}}\mathbf{1} + \mathbf{AB0} = \overline{\mathbf{AB}}$; (b) Concatenated CCNOT gates.

In Fig. 1.5 (b), the output of the first $\oplus$ in the third line is the controlled input to the second $\oplus$. The final output is therefore

$$\begin{aligned} \mathbf{C}^{\neg\neg\neg\neg} &= \overline{\mathbf{AB}}(\overline{\mathbf{AB}}\mathbf{C} + \mathbf{AB}\overline{\mathbf{C}}) + \mathbf{AB}\overline{(\overline{\mathbf{AB}}\mathbf{C} + \mathbf{AB}\overline{\mathbf{C}})} \\ &= \overline{\mathbf{AB}}\mathbf{C} + \mathbf{AB}(\mathbf{AB} + \overline{\mathbf{C}})(\overline{\mathbf{AB}} + \mathbf{C}) \\ &= (\overline{\mathbf{AB}} + \mathbf{AB})\mathbf{C} \;=\; \mathbf{C}, \end{aligned} \tag{1.6}$$

where we used distributivity: $\mathbf{A}(\mathbf{B} + \mathbf{C}) = \mathbf{AB} + \mathbf{AC}$ and De Morgan's laws: $\overline{\mathbf{AB}} = \overline{\mathbf{A}} + \overline{\mathbf{B}}$, $\overline{(\mathbf{A} + \mathbf{B})} = \overline{\mathbf{A}}\,\overline{\mathbf{B}}$, which hold in any Boolean algebra. The use of such general Boolean expressions is not only more general—such expressions also admit multiple-valued logic [Hurst, 1984]—but often also shorter and more elegant than handling 0–1 values and modulo 2 operations.

A formal advantage of NOT, CNOT, and CCNOT gates is that NOT gate can be represented—see Eq. (1.5)—by a matrix acting on an input column vector and giving an output column vector, as opposed to truth tables (cf. NAND Karnaugh map [Helms, 1983] in Fig. 1.3), which do not form a matrix algebra. Thus CNOT and CCNOT have particularly simple forms. Let us consider CCNOT acting on three binary inputs $\mathbf{A}, \mathbf{B}, \mathbf{C}$, forming the set $\{\{0,0,0\},\{0,0,1\},\ldots,\{1,1,1\}\}$ of $2^3 = 8$ elements. Let us denote $\{\mathbf{A}, \mathbf{B}, \mathbf{C}\}$ by 8×1 (8 rows and 1 column) matrices: $\{0,0,0\}$ will have 1 in the first row and 0s in all the others, $\{0,0,1\}$ will have 1 in the second row and 0s in all the others, etc. CCNOT has the following 8×8 matrix, which we denote by $\widehat{\text{CCNOT}}$:

$$\widehat{\text{CCNOT}}\, a = \begin{bmatrix} 1 & 0 & 0 & 0 & 0 & 0 & 0 & 0 \\ 0 & 1 & 0 & 0 & 0 & 0 & 0 & 0 \\ 0 & 0 & 1 & 0 & 0 & 0 & 0 & 0 \\ 0 & 0 & 0 & 1 & 0 & 0 & 0 & 0 \\ 0 & 0 & 0 & 0 & 1 & 0 & 0 & 0 \\ 0 & 0 & 0 & 0 & 0 & 1 & 0 & 0 \\ 0 & 0 & 0 & 0 & 0 & 0 & 0 & 1 \\ 0 & 0 & 0 & 0 & 0 & 0 & 1 & 0 \end{bmatrix} \begin{bmatrix} 0 \\ 0 \\ 0 \\ 0 \\ 0 \\ 0 \\ 0 \\ 1 \end{bmatrix} = \begin{bmatrix} 0 \\ 0 \\ 0 \\ 0 \\ 0 \\ 0 \\ 1 \\ 0 \end{bmatrix} = b \quad (1.7)$$

which is here shown acting on $a = \{1,1,1\}$ and mapping it into $b = \{1,1,0\}$. The reversibility is obvious from $\widehat{\text{CCNOT}}\ b = a$ and/or $\widehat{\text{CCNOT}} \cdot \widehat{\text{CCNOT}} = \mathbb{1}$, where $\mathbb{1}$ is the unit matrix.

1.8 Quantum Bits: Qubits

We mentioned above that reversible computer and quantum computer theories have many characteristics in common. The main characteristic is the reversibility of the gates. Classical reversible computers are reversible so as to avoid energy dissipation, and they use reversible classical gates, binary or multivalued, but in any case with definite values at each stage of calculation. Quantum computers are reversible because of their quantum nature, and so are their gates, known as *quantum gates.* The units of information that the latter gates process— *qu*antum *bits*, or *qubits*—have no definite values at some intermediate stages of calculation. Physically, a qubit is a two-level, two-dimensional system that can be represented by a vector from a two-dimensional space. Since any vector in two-dimensional space can be written as a linear combination of two perpendicular vectors determined by two measuring devices, or detectors, there are infinitely many vectors that can determine the state of a qubit before a measurement.

To see how this works, let us look at the well known polarization experiment shown in Fig. 1.6. The figure shows unpolarized photons leaving the source S one by one. A photon is a combination of the particle and wave aspects of light. The quantum, or the *particle* aspect of the photons in the experiment, can transfer its energy to other systems so that we can read off the energy with detector D. This Planck energy $E = h\nu$ (where h is Planck's constant [6.63×10^{-34} J·s] and ν is the frequency of the light given by its wave representation) can be transferred only in discrete amounts to a detector: 0 (no "click") and $h\nu$ (one click), analogous to computer bits 0 and 1. But the analogy is not straightforward, since the wave aspect of a photon is partly responsible for its properties. We can also represent the photon by means of the electric-field vector

$$\begin{aligned} \mathbf{E}(\mathbf{r},t) &= \mathbf{E}_0 e^{i(\mathbf{k}\cdot\mathbf{r}-\omega t-\epsilon)} \\ &= \mathbf{E}_0(\cos(\mathbf{k}\cdot\mathbf{r}-\omega t-\epsilon)+i\,\sin(\mathbf{k}\cdot\mathbf{r}-\omega t-\epsilon)), \qquad (1.8) \end{aligned}$$

where $\mathbf{E}_0$ is the amplitude of the wave, $\mathbf{k} = \mathbf{k}_0\frac{2\pi}{\lambda}$, $\mathbf{k}_0$ is the direction in which the wave propagates ($\mathbf{k}_0 \perp \mathbf{E}_0$), λ is the wavelength ($\lambda\nu = c$ = speed of light), $\omega = 2\pi\nu$, and ϵ is a *phase shift.* Its components vary sinusoidally with time, but we have no medium that carries the "variations"—waves are nothing but a description of photon's properties.

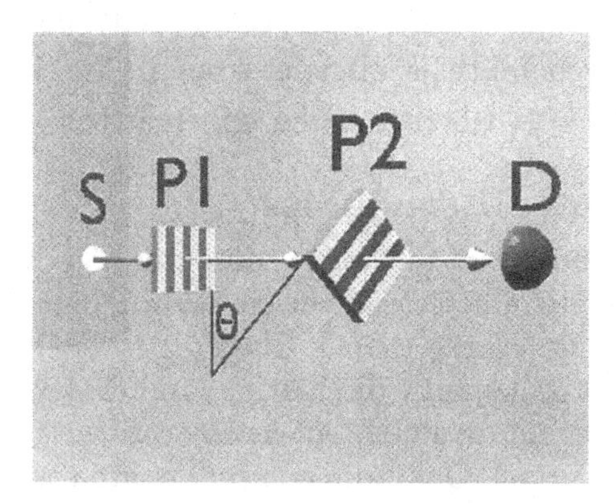

Figure 1.6. Polarization experiment. Photons—one by one—come from the source S, and some pass through polarizers P1 and P2 to be detected by the detector D; P2 is rotated by Θ with respect to P1—here shown in an Escher-like way.

In the above experiment, photons leave the source S "without properties" in the sense that the orientation of $\mathbf{E}_0$ in time is not determined.

By using the (linear) polarizer P1, we impose the property of "vertical polarization" upon photons that pass through P1 (50% of incoming photons). We describe this property by choosing $E_{0y} = 0$ and $E_{0z} = 0$ so as to have $\mathbf{E}_0 = E_{0x}\mathbf{j}$, i.e., $\mathbf{E}_0$ stays in the xz-plane (photons move along the z-axis). We also say that we *prepare* a photon with vertical polarization by letting it pass through P1.

We can *measure* the preparation with the help of P2 oriented parallel to P1 ($\theta = 0$): all photons pass through. When P2 is rotated by $\theta = \pm\frac{\pi}{2}$, no photons pass through. So, Jones vectors (a) and (c) of Fig. 1.7 correspond to the two possible values of classical bits 0 and 1. But there are infinitely many intermediate positions that do not—one of them is the vector (b) of Fig. 1.7. The average value of energy flow carried by an electromagnetic wave is given by the average value $\langle S \rangle = c\epsilon_0 \langle \mathbf{EE}^* \rangle = \frac{c\epsilon_0}{2} E_0^2$ of the Poynting vector $\mathbf{S}$ (see p. 21), where * means the complex conjugate and $\epsilon_0 = 8.854 \times 10t^{-12}\ F/m$ is the permittivity of vacuum [Saleh and Teich, 1991]. The irradiance, which is $I(0) = \langle S \rangle$ coming out from P1, reduces to

$$I(\theta) = \frac{c\epsilon_0}{2}(E_0 \cos\theta)^2 = I(0)\cos^2\theta \tag{1.9}$$

after passing through P2. Eq. (1.9) is called the *Malus law.* Let us suppose 200 photons come out from the source S and let us suppose P2 is rotated by $\theta = 30°$. Then 100 photons on average will pass through P1, and we will register $100\cos^2 30° = 75$ *clicks* with detector D. We could detect the other 25 photons if we used a birefringent crystal [Saleh and Teich, 1991] for P2: 75 photons would then end up in D and 25 in detector $\mathrm{D}^{\perp}$ detecting perpendicularly polarized photons.

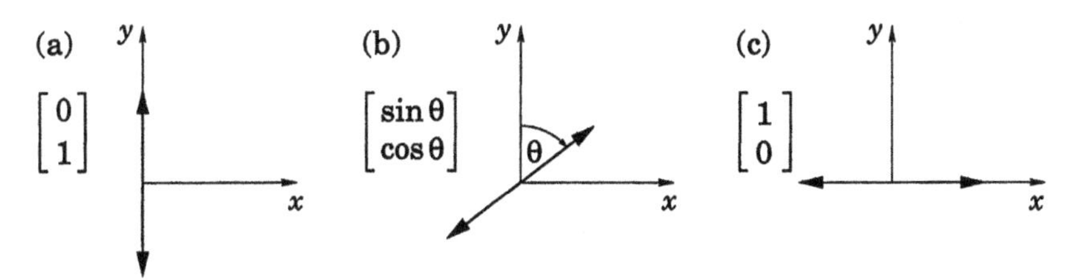

Figure 1.7. Jones vectors: (a) linearly polarized wave in y direction; (b) in a direction making an angle θ with the x-axis; (c) in the x direction.

The inescapable conclusion from this experiment is that qubits—here linear photon polarizations along the x- and y-axes—are not fixed properties attached to their carriers (photons) and/or devices (polarizers). Of all the photons polarized along x and coming from P1, polarizer P2 lets some of them through and blocks others. However, when $\theta \neq 0$, there is no way to know in advance which of the photons will be let through and which not. All we know is their probability of passing through the polarizer. Hence, a single device as shown in Fig. 1.6 might have—as opposed to a transistor—many possible outputs. How can we make such a probabilistic device calculate anything? The answer lies in superposition and entanglement.

1.9 Flying Qubits and Circular Polarization

In Sec. 3.1.6 (p. 146) we will use polarization states of photons to transfer a state of one atom to another. When building a *quantum network* for computation and communication, photons that carry a quantum state are called *flying qubits*, as opposed to atoms, which are then called *stationary qubits*. To be able to calculate the behavior and properties of flying qubits, we have to use a "more quantum" description of photons than in Sec. 1.8. This also requires that we start with a different type of polarization—circular polarization. In the end we can use it to define linear polarization from Sec. 1.8.

Classically, polarization is the direction of a transverse electric field $\mathbf{E}$ at a particular moment. Hence, it is a vector perpendicular to the direction of the propagation wave vector $\mathbf{k}$. If $\mathbf{k}$ is oriented along the z-axis, the electric-field vector components are

$$E_x = E_{0x}e^{kz-i\omega t} \qquad \text{and} \qquad E_y = E_{0y}e^{kz-i\omega t+\delta}. \tag{1.10}$$

It is obvious that the projection of the real part of this vector onto the xy-plane is an ellipse. For $E_{0x} = E_{0y}$ and $\delta = \pi/2$ we have the right circular polarization and for $\delta = -\pi/2$ the left circular polarization. Their Jones vectors are shown in Fig. 1.8. $|R\rangle$ and $|L\rangle$ are the so-called *ket* vectors belonging to Dirac's *bra–ket* notation for writing down state functions and state vectors. The notation is widely used in quantum physics, and we will elaborate on it in Sec. 1.11.

Quantum mechanically, a circularly polarized photon has a well-defined angular momentum along $\mathbf{k}$. The angular momentum vector of an electromagnetic radiation field (for example, the field of a laser beam) is [Messiah, 1965]

$$\mathbf{j}_p = \frac{c}{4\pi}\int_V \left[\mathbf{r} \times (\mathbf{E}_\perp \times \mathbf{B})\right] dV, \tag{1.11}$$

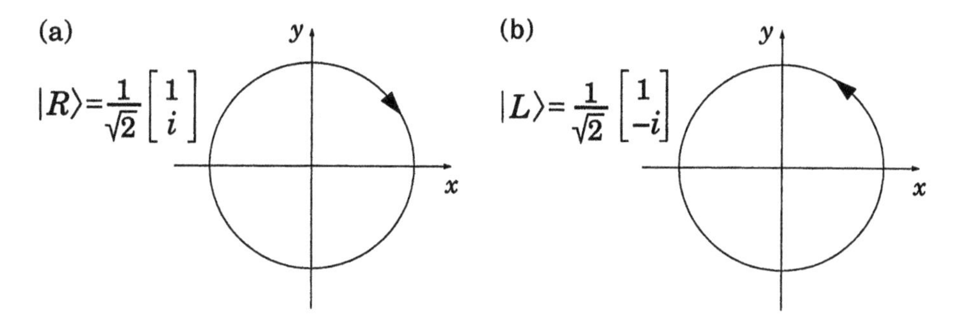

Figure 1.8. Jones vectors: (a) Right-hand circularly polarized light; (b) Left-hand circularly polarized light

where **E** and **B** are electrical and magnetic radiation fields and $\mathbf{E}_\perp$ is the *transversal*[6] part of the electric field **E**. A photon is a *carrier* (corpuscle, particle) of the electromagnetic radiation field. We assume that the photon is in either the $|R\rangle$ or or $|L\rangle$ state. It is straightforward but tedious to show that [Messiah, 1965]

$$\frac{\mathbf{k}\cdot\mathbf{j}_p}{k}|R\rangle = +|R\rangle, \qquad \frac{\mathbf{k}\cdot\mathbf{j}_p}{k}|L\rangle = -|L\rangle, \tag{1.12}$$

i.e., that the component of $\mathbf{j}_p$ along **k** is equal to either +1 or -1. Hence, $\mathbf{j}_p$ is the total photon angular momentum here,[7] $|R\rangle$, $|L\rangle$ are the eigenvectors of $\mathbf{k}\cdot\mathbf{j}_p/k$, and $m_{j_p} = \pm 1$ the eigenvalues.

Now, linear polarization (Fig. 1.7, p. 19) emerges as a superposition of these eigenvectors:

$$\begin{bmatrix} 0 \\ 1 \end{bmatrix} = \frac{i}{\sqrt{2}}(|L\rangle - |R\rangle) \quad \text{and} \quad \begin{bmatrix} 1 \\ 0 \end{bmatrix} = \frac{1}{\sqrt{2}}(|L\rangle + |R\rangle). \tag{1.13}$$

[6] Any vector field **F** can be considered to be a superposition of a *divergenceless field* $\mathbf{F}_\perp$ ($\mathrm{div}\mathbf{F}_\perp = 0$, $\mathrm{curl}\mathbf{F}_\perp = \mathrm{curl}\mathbf{F}$) and an *irrotational field* $\mathbf{F}_\parallel$ ($\mathrm{curl}\mathbf{F}_\parallel = 0$, $\mathrm{div}\mathbf{F}_\parallel = \mathrm{div}\mathbf{F}$). For square integrable vector fields—and we will always assume that our fields are square integrable—this decomposition is unique, and we speak of a *longitudinal* and *transversal* part of **F**, respectively. Maxwell's equations require $\mathbf{B}_\parallel = 0$, and $\mathbf{E}_\parallel$ is the electrostatic field in the presence of charge. In the absence of charge, $\mathbf{E}_\parallel = 0$ and $\mathbf{E} = \mathbf{E}_\perp$. This is our case, and $\mathbf{j}_p$, given by Eq. (1.11), reads $\frac{c}{4\pi}\int_V[\mathbf{r}\times(\mathbf{E}\times\mathbf{B})]dV$. We recognize that $\mathbf{E}\times\mathbf{B} = \mathbf{S}/c^2$, where **S** is the *Poynting vector*. It is a density of electromagnetic momentum and therefore $\mathbf{r}\times\mathbf{S}/c^2$ is the density of the angular electromagnetic momentum, and $\mathbf{j}_p$ is the angular momentum [Jackson, 1967].

[7] Here the total angular momentum coincides with the spin (circular polarization) angular momentum. However, one can impose an azimuthal phase dependence $\exp(-il\phi)$, where l is any integer, on a photon, thus giving it an orbital momentum as well [Padgett et al., 2004]. Then the total angular momentum becomes a combination of these two angular momentums, as with other quantum particles.

Physically, we change polarization by a *phase retarder*, which is a plate that shifts the phase and has the following matrix representation:

$$\begin{matrix} E_{0x}e^{i\phi_x} \to E_{0x}e^{i\phi_x+\delta_x} \\ E_{0y}e^{i\phi_y} \to E_{0y}e^{i\phi_y+\delta_y} \end{matrix} \quad \Leftrightarrow \quad \begin{bmatrix} e^{i\delta_x} & 0 \\ 0 & e^{i\delta_y} \end{bmatrix}. \tag{1.14}$$

The so-called *quarter-wave plate* (QWP) and *half-wave plate* (HWP) are the most often used phase retarders:

$$\text{QWP}: \quad e^{i\pi/4}\begin{bmatrix} 1 & 0 \\ 0 & i \end{bmatrix}, \qquad \text{HWP}: \quad e^{i\pi/2}\begin{bmatrix} 1 & 0 \\ 0 & -1 \end{bmatrix}, \tag{1.15}$$

where the y-axis is chosen as the "slow axis."

A QWP changes a linear polarization at $-45°$ into a right-hand circular polarization, and a left-hand circular polarization into a linear polarization at $+45°$:

$$\begin{bmatrix} 1 & 0 \\ 0 & i \end{bmatrix}\begin{bmatrix} 1 \\ 1 \end{bmatrix} = \begin{bmatrix} 1 \\ i \end{bmatrix}, \qquad \begin{bmatrix} 1 & 0 \\ 0 & i \end{bmatrix}\begin{bmatrix} 1 \\ -i \end{bmatrix} = \begin{bmatrix} 1 \\ 1 \end{bmatrix}. \tag{1.16}$$

An HWP changes a linear polarization at $-45°$ into a linear polarization at $+45°$, and a right-hand circular polarization into a left-hand one:

$$\begin{bmatrix} 1 & 0 \\ 0 & -1 \end{bmatrix}\begin{bmatrix} 1 \\ 1 \end{bmatrix} = \begin{bmatrix} 1 \\ -1 \end{bmatrix}, \qquad \begin{bmatrix} 1 & 0 \\ 0 & -1 \end{bmatrix}\begin{bmatrix} 1 \\ -i \end{bmatrix} = \begin{bmatrix} 1 \\ i \end{bmatrix}. \tag{1.17}$$

1.10 Superposition of Qubits

In the previous section, we used the term *qubit* (*quantum bit*) for the information ascribed to two polarization states of photons that are interacting with devices (polarizers; environment). In doing so, we only used one part of the electric-field description of photons, namely, amplitude, while neglecting the other: phase. We could do this because we used only one photon path for the photons. When we use more than one path and combine them, we arrive at their *superposition*, which essentially relies on their phase difference. We shall first consider phase description alone and come to a combination of phase and polarization later on.

We have already mentioned that, physically, a logic gate in a classical computer is a transistor circuit that allows voltages to pass through based on simple Boolean algebraic (logic) rules applied to its inputs. Thus, gates for the same Boolean operation differ according to their physical implementations; for example, the PMOS classical NAND gate that Richard Feynman presented in his seminal paper [Feynman, 1985] differs from today's CMOS NAND gate (presented in Fig. 1.3, p. 13). In the literature, a logic gate may be represented by its algebraic Boolean representation or by more complex electronic representations that include delays and other parametric information.

A quantum (logic) gate is not easy to define physically because there are many possible physical realizations of gates and ultimately quantum computers under consideration. We also cannot have truth tables for an algebraic representation of qubits. Therefore, the usual approach is to define a gate physically for a considered implementation and algebraically by means of input–output transformation rules—*before* measurement—for qubits. It turns out that these rules can be given by unitary operators, usually with a matrix representation.

To eventually arrive at quantum logic gates, let us first consider a simple experiment consisting of a photon splitting its path at a beam splitter, as shown in Fig. 1.9. We denote the two possible incoming paths and also the corresponding states of the photon moving along them by $|0\rangle$ and $|1\rangle$. These correspond to electric-field vectors along the incoming paths. So the photon arrives either from above and has the state described by $|0\rangle$ or from below in state $|1\rangle$. These possibilities correspond to an electric field describing a wave of frequency ν that propagates along either path 0 or path 1. Actually, we make no distinction between a wave function and a state vector.

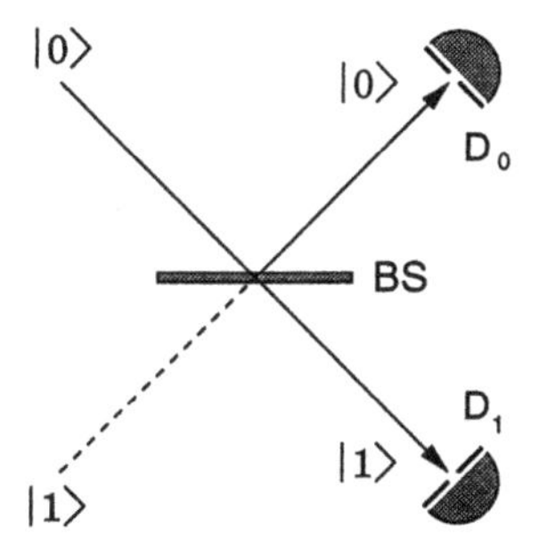

Figure 1.9. Photon at a beam splitter as a $\sqrt{\text{NOT}}$ gate (*before* the collapse of the wave packet in one of the detectors D_i, $i = 0, 1$).

The photon can either go through or be reflected from the beam splitter. Let us take the case of photon $|0\rangle$ coming in. If it passes through, its field vector will remain unchanged and we will only change its name because it is now below the beam splitter BS. But because it passes through BS with only 50% probability, we multiply its ket by $1/\sqrt{2}$. On the other hand, a vector field reflected from BS undergoes a phase shift $\pi/2$ with respect to the field that passes through it. (See [Degiorgio, 1980], where one must assume that the lower incoming beam does not contain a photon.) This phase shift corresponds to multiplying the ket by $e^{i\pi/2} = i$; therefore, the reflected photon will be described by $(1/\sqrt{2})i|0\rangle$. Hence, *before* we detect which outgoing path the photon took—by registering a "click" in either D_0 or D_1—we describe its state by the following *superposition* of paths:

$$|\text{out}\rangle = \frac{1}{\sqrt{2}}(|1\rangle + i|0\rangle). \tag{1.18}$$

This photon behavior makes a beam splitter the so-called $\sqrt{\text{NOT}}$ gate—up to a phase shift (see Section 1.14, p. 29).

To see how the detectors will register the states of our photon, i.e., our qubit, we will review the following basics of quantum theory [Messiah, 1965].

1.11 Bra–Ket Qubit Formalism

All the vectors of states belong to a *vector space.* One such kind of vector is *kets,* which we have already introduced. Vectors that are *dual* to kets are *bra* vectors. The bra vector $\langle\Psi|$ is the conjugate of the ket $|\Psi\rangle$:

$$\langle\Psi| = |\Psi\rangle^*. \tag{1.19}$$

Take, for example, the electric-field ket vector

$$|0\rangle = \mathbf{E}_0 e^{i(kx-\omega t-\epsilon)} \tag{1.20}$$

for our polarization measurement above (after the photon passes through P1). Its bra is

$$\langle 0| = |0\rangle^* = \mathbf{E}_0 e^{-i(kx-\omega t-\epsilon)}. \tag{1.21}$$

Hence, for

$$|\psi\rangle = \alpha|0\rangle + \beta|1\rangle, \tag{1.22}$$

we have

$$\langle\psi| = \alpha^*\langle 0| + \beta^*\langle 1|. \tag{1.23}$$

In the matrix representation, $|0\rangle$, $|1\rangle$, the bra given by Eq. (1.22) and the ket given by Eq. (1.23) read

$$|0\rangle = \begin{bmatrix} 0 \\ 1 \end{bmatrix}; \quad |1\rangle = \begin{bmatrix} 1 \\ 0 \end{bmatrix}; \quad |\psi\rangle = \begin{bmatrix} \alpha \\ \beta \end{bmatrix}; \quad \langle\psi| = [\,\alpha^* \; \beta^*\,]. \tag{1.24}$$

Hence, the conjugation relations between vectors correspond to Hermitian conjugation between matrices (transposition + conjugation).

If $|\psi_1\rangle$ and $|\psi_2\rangle$ are two n-dimensional kets, any linear combination of them,

$$\lambda_1|\psi_1\rangle + \lambda_2|\psi_2\rangle, \tag{1.25}$$

where λ_i, $i = 1, 2$ are complex numbers, is again a ket, namely, $|\lambda_1\psi_1 + \lambda_2\psi_2\rangle$. In other words, the vector space containing our kets and bras is *linear*.

The scalar product of the ket $|\psi_1\rangle$ and the bra $\langle\psi_2|$ is a complex number $\langle\psi_2|\psi_1\rangle$, called a *bracket*, which is determined by the fact that our vectors are state vectors—electric-field vectors, wave functions—from a complex linear vector space with an orthonormal basis (a set of mutually orthogonal vectors of unit length by means of which any vector from the space can be expressed). If $|\psi_i\rangle$, $i = 1, 2$, forms an orthonormal basis, we will have $\langle\psi_i|\psi_j\rangle = \delta_{ij}$, $i, j = 1, 2$, where $\delta_{ij} = 1$ for $i = j$ and $\delta_{ij} = 0$ for $i \neq j$ (δ_{ij} is called the *delta function*). Thus for $|0\rangle$ and $|1\rangle$ of length 1 we have

$$\langle 0|0\rangle = 1, \quad \langle 0|1\rangle = 0, \quad \langle 1|0\rangle = 0, \quad \langle 1|1\rangle = 1, \tag{1.26}$$

where the brackets correspond to the inner product $\mathbf{a}\cdot\mathbf{b} = ab\cos\angle(\mathbf{a}, \mathbf{b})$, in the classical 2-dimensional vector space. For our polarization experiment with the vector $\mathbf{E}_0$ of length 1, we have for $|0\rangle$ given by Eq. (1.21):

$$|0\rangle = \mathbf{i}\, e^{i(kz-\omega t-\epsilon)} \tag{1.27}$$

and for $|0\rangle$ (the polarizer oriented along the y-axis):

$$|1\rangle = \mathbf{j}\, e^{i(kz-\omega t-\epsilon)} \tag{1.28}$$

Hence, for example, $\langle 0|0\rangle = \mathbf{i}\cdot\mathbf{i}\, e^{i(kz-\omega t-\epsilon)} e^{-i(kz-\omega t-\epsilon)} = 1$, $\langle 0|1\rangle = \mathbf{i}\cdot\mathbf{j}\, e^{i(kz-\omega t-\epsilon)}\, e^{-i(kz-\omega t-\epsilon)} = 0$.

To summarize, the scalar product of $|\psi_1\rangle$ and $\langle\psi_2|$ is the complex conjugate of the scalar product of $|\psi_2\rangle$ and $\langle\psi_1|$

$$\langle\psi_2|\psi_1\rangle = \langle\psi_1|\psi_2\rangle^*, \tag{1.29}$$

and the scalar product of $|\psi\rangle$ and $\langle\phi|$ is linear with respect to $|\psi\rangle$

$$\langle\phi|\lambda_1\psi_1 + \lambda_2\psi_2\rangle = \lambda_1\langle\phi|\psi_1\rangle + \lambda_2\langle\phi|\psi_2\rangle. \tag{1.30}$$

The *norm* of $|\psi\rangle$, $N_\psi \stackrel{\text{def}}{=} \langle\psi|\psi\rangle$, is a real nonnegative number:

$$\langle\psi|\psi\rangle \geq 0. \tag{1.31}$$

For quantum computing, we only need to use finite dimensional spaces. Since every finite-dimensional complex vector space with scalar product is (a) *complete* (every subspace is closed) and, in our case, (b) *separable* (has at most countable orthonormal basis) [Weidman, 1980], it turns out that our kets and bras belong to a Hilbert space by definition.

1.12 Operators

In a Hilbert space, we can express any operator and therefore any symbolic representation of a gate by means of bra and ket vectors. For instance, the NOT gate can be expressed as the following operator:

$$\begin{aligned}\widehat{\text{NOT}} &= \begin{bmatrix} 0 & 1 \\ 1 & 0 \end{bmatrix} = \begin{bmatrix} 0 & 1 \\ 0 & 0 \end{bmatrix} + \begin{bmatrix} 0 & 0 \\ 1 & 0 \end{bmatrix} \\ &= \begin{bmatrix} 1 \\ 0 \end{bmatrix} [0\ 1] + \begin{bmatrix} 0 \\ 1 \end{bmatrix} [1\ 0] = |0\rangle\langle 1| + |1\rangle\langle 0|,\end{aligned} \tag{1.32}$$

and a general two-dimensional operator as

$$\begin{bmatrix} \alpha_{00} & \alpha_{01} \\ \alpha_{10} & \alpha_{11} \end{bmatrix} = \alpha_{00}|0\rangle\langle 0| + \alpha_{01}|0\rangle\langle 1| + \alpha_{10}|1\rangle\langle 0| + \alpha_{11}|1\rangle\langle 1|. \tag{1.33}$$

These equations suggest a general meaning for an operator in Hilbert space. A linear operator A describes a linear correspondence between vectors—say, $|\psi\rangle$ and $|\phi\rangle$ (single column matrices)—in Hilbert space:

$$|\phi\rangle = \hat{A}|\psi\rangle, \tag{1.34}$$

where a "hat" over a symbol means it is an operator and/or a matrix; we will usually omit the hat whenever there is no risk of confusion.

The so-called *adjoint* or *Hermitian conjugate* operator $A^\dagger$ of A,

$$\begin{bmatrix} a_{11} & \cdots & a_{1n} \\ \vdots & \ddots & \vdots \\ a_{n1} & \cdots & a_{nn} \end{bmatrix}^\dagger = \begin{bmatrix} a_{11}^* & \cdots & a_{1n}^* \\ \vdots & \ddots & \vdots \\ a_{n1}^* & \cdots & a_{nn}^* \end{bmatrix}^{\mathrm{T}} = \begin{bmatrix} a_{11}^* & \cdots & a_{n1}^* \\ \vdots & \ddots & \vdots \\ a_{1n}^* & \cdots & a_{nn}^* \end{bmatrix}, \tag{1.35}$$

describes a dual correspondence between bra vectors $\langle\psi|$ and $\langle\phi|$ (single-row matrices):

$$\langle\phi| = \langle\psi|A^\dagger.$$

A complex number a is an *eigenvalue* of the linear operator A and the ket $|\psi\rangle$, an *eigenket* (*eigenvector, eigenfunction*) associated with a if

$$A|\psi\rangle = a|\psi\rangle. \tag{1.36}$$

If a linear operator H is its own adjoint, it is called *Hermitian*:

$$H = H^\dagger. \tag{1.37}$$

For example, the operator $|\psi\rangle\langle\psi|$ is a Hermitian operator.

If an operator U is the inverse of its own adjoint, it is called *unitary*:

$$U\,U^\dagger = U^\dagger\,U = \mathbb{1}. \tag{1.38}$$

It should be noted here that a product of two unitary operators is a unitary operator while a product of two Hermitian operators is not (in general) a Hermitian operator.

For a measurement, aside from particular special operators, the most important operators are linear operators called *projection operators* or simply *projectors*, P on a linear subspace S:

$$P_S|\psi\rangle = |\psi_S\rangle \tag{1.39}$$

P_S is a Hermitian operator and satisfies the operator equation

$$P_S^2 = P_S. \tag{1.40}$$

Also, any Hermitian operator satisfying Eq. (1.40) is a projector.

1.13 Detecting Qubits

A quantum measurement, i.e., a detection of qubits, is intrinsically probabilistic. A detection consists of "clicks," i.e., of finding a system in one of two possible states, either $|0\rangle$ or $|1\rangle$; but before a detections takes place, we deal only with probability amplitudes and probabilities.

Thus, the probability of finding a quantum system—which was in the dynamical state $|\psi\rangle$ (normalised to unity) just *before* the measurement was carried out—to be in the state $|\phi\rangle$ is equal to

$$|\langle\phi|\psi\rangle|^2. \tag{1.41}$$

More generally, the probability of finding the system in any one of the states of a subspace S_D is equal to the average (mean value) of the projector P_D on S_D:

$$\langle P_D\rangle = \langle\psi|P_D|\psi\rangle, \tag{1.42}$$

and the mean value of any function $F(A)$ of a given physical quantity is

$$\langle F(A)\rangle = \langle\psi|F(A)|\psi\rangle. \tag{1.43}$$

When the dynamical state mentioned in connection with Eq. (1.41) is known completely, it can be represented by a unique ket. When it is known only up to probabilities p_i, $i = 1, \ldots, n$, of a system being in states given by kets $|\psi_i\rangle$, then the dynamical state is represented by a *statistical mixture* of kets. In this case, the mean value given by Eq. (1.43) is given by

$$\langle F(A)\rangle = \sum_n p_n \langle\psi_n|F(A)|\psi_n\rangle. \tag{1.44}$$

A particularly efficient description of statistical mixtures can be achieved by means of the so-called *density operator*

$$\rho = \sum_n |\psi_n\rangle p_n \langle\psi_n|, \tag{1.45}$$

with the help of which we can define the probability of finding the system in the state of $|\psi\rangle$ as

$$w_\psi = \mathrm{Tr}(\rho|\psi\rangle\langle\psi|) = \langle\psi|\rho|\psi\rangle, \tag{1.46}$$

and the mean value, Eq. (1.43), as

$$\langle F(A)\rangle = \mathrm{Tr}\rho F(A). \tag{1.47}$$

When the dynamical state $|\psi\rangle$ of a system is completely known, as in the earlier case above, it is called a *pure state*. We can represent it as a statistical mixture having $|\psi\rangle$ as its only element. The density operator of a pure state is the projector

$$\rho_\psi = |\psi\rangle\langle\psi|, \tag{1.48}$$

and we can easily verify property (1.40):

$$\rho_\psi^2 = \rho_\psi. \tag{1.49}$$

Let us now consider measurements by detectors D_0 and D_1 in the beam splitter experiment presented in Sec. 1.10 (p. 22) and shown in Fig. 1.9 (p. 23). Detection of a photon by detector D_0 according to Eq. (1.41) occurs with the probability

$$|\langle 0|\mathrm{out}\rangle|^2 = \frac{1}{2}, \tag{1.50}$$

where we made use of Eqs. (1.18) and (1.26). The same probability we obtain for D_1. As in Sec. 1.8 this means that we will always detect a photon by either D_0 or D_1 with a probability of 50%. The density operator of the function $|\mathrm{out}\rangle$ given by Eq. (1.18) is

$$\rho_{\mathrm{out}} = \frac{1}{2}|0\rangle\langle 0| + \frac{1}{2}|1\rangle\langle 1|. \tag{1.51}$$

At first glance, this detection of a photon looks like a coin-flip measurement. However, the "flips" are not classical, as we will see in the next section.

1.14 Quantum Gates and Circuits

The mathematical description of photon behavior at a beam splitter by means of electric-field operators shows that we can reverse the spreading of photon waves in time. Let us see how this process works on a second beam splitter, shown in Fig. 1.10 within a device called a *Mach–Zehnder interferometer.*

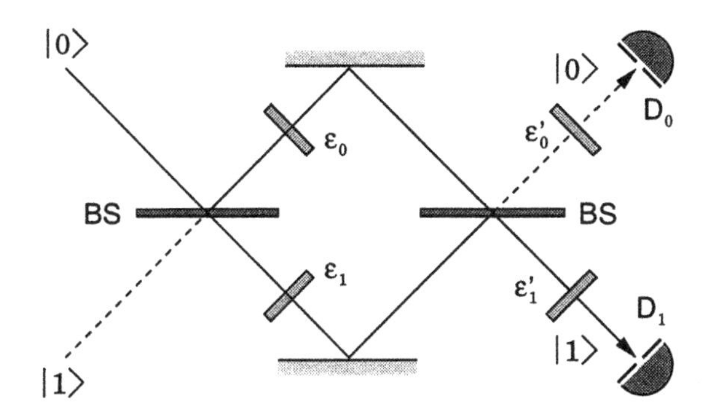

Figure 1.10. Mach–Zehnder interferometer implementation of $\sqrt{\text{NOT}}\sqrt{\text{NOT}} = \text{NOT}$ gate: An incoming $|0\rangle$ ($|1\rangle$) photon will always end up in the D_1 (D_0) detector.

The path to the second beam splitter (BS) from above is described by $(i/\sqrt{2})|0\rangle$ and from below by $(1/\sqrt{2})|1\rangle$. Here we can simply reverse the process we had at the first beam splitter in Sec. 1.10 as follows. The two paths superpose at the beam splitter, so the the upper outgoing path is described by

$$\frac{1}{\sqrt{2}}(\frac{1}{\sqrt{2}}|0\rangle + i\frac{i}{\sqrt{2}}|0\rangle) = 0 \tag{1.52}$$

and the lower one by

$$\frac{1}{\sqrt{2}}(\frac{i}{\sqrt{2}}|1\rangle + i\frac{1}{\sqrt{2}}|1\rangle) = i|1\rangle \tag{1.53}$$

when the phase shifters ϵ_0 and ϵ_1 are set at $\epsilon_0 = 0$ and $\epsilon_1 = 0$, i.e., they do not play a role. In other words, the process at the second beam splitter is just a reverse image of the process at the first one.

Hence, following Eq. (1.41), the probability of detecting the photon by D_0 is 0, and the probability of detecting it by D_1 is $\left|\langle 1|(-i)i|1\rangle\right|^2 = 1$. In other words, we have a NOT gate—up to an irrelevant phase shift ($e^{i\pi/2} = i$), which we can eliminate by adding a $-\pi/2$ phase shifter at both exits.

Using matrix representation, we could describe the action of the Mach–Zehnder interferometer, when phase shifters ϵ_0' and ϵ_1' are set at $-\pi/2$ and when ϵ_0 and ϵ_1 are taken out, as follows:

$$\begin{bmatrix} e^{-i\frac{\pi}{2}} & 0 \\ 0 & e^{-i\frac{\pi}{2}} \end{bmatrix} \frac{1}{\sqrt{2}} \begin{bmatrix} 1 & i \\ i & 1 \end{bmatrix} \frac{1}{\sqrt{2}} \begin{bmatrix} 1 & i \\ i & 1 \end{bmatrix} = \begin{bmatrix} 0 & 1 \\ 1 & 0 \end{bmatrix}. \tag{1.54}$$

This is why, in Section 1.10, we called a beam splitter a $\sqrt{\text{NOT}}$ gate—up to a phase shift: its repeated use switches the paths and acts as a NOT gate with an irrelevant phase shift. A literal $\sqrt{\text{NOT}}$ gate, whose repeated application gives a NOT gate without a phase shift, is given as a beam splitter with phase shifters at the exits set at $-\pi/4$. In our case, this means ϵ_i, ϵ_i', $i = 0, 1$ set to $-\pi/4$. The transformation then reads:

$$\left(\sqrt{\text{NOT}}\right)^2 = \left(\begin{bmatrix} e^{-i\frac{\pi}{4}} & 0 \\ 0 & e^{-i\frac{\pi}{4}} \end{bmatrix} \frac{1}{\sqrt{2}} \begin{bmatrix} 1 & i \\ i & 1 \end{bmatrix} \right)^2 = \begin{bmatrix} 0 & 1 \\ 1 & 0 \end{bmatrix}. \tag{1.55}$$

Hence we obtain

$$\sqrt{\text{NOT}} = \frac{1-i}{2} \begin{bmatrix} 1 & i \\ i & 1 \end{bmatrix}. \tag{1.56}$$

Such one-qubit gates we describe by the quantum circuit diagrams shown in Fig. 1.11. By a *quantum circuit* we understand a device con-

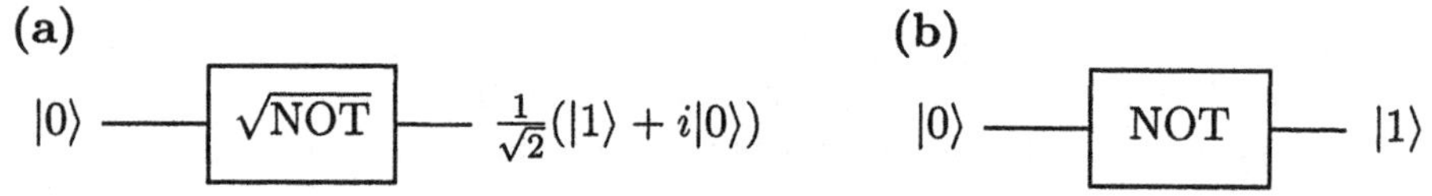

(a) $|0\rangle$ — $\sqrt{\text{NOT}}$ — $\frac{1}{\sqrt{2}}(|1\rangle + i|0\rangle)$

(b) $|0\rangle$ — NOT — $|1\rangle$

Figure 1.11. (a) Quantum $\sqrt{\text{NOT}}$ gate—in this section given by Eqs. (1.55) and (1.56); (b) quantum NOT gate—in this section obtained by concatenation of two consecutive $\sqrt{\text{NOT}}$ gates, per Eq. (1.54).

sisting of quantum gates and arranged according to steps in which the gates process qubits in time; time is assumed to run from the left- to the right-hand side of a diagram. The number of gates a circuit consists of is called its *size*. The size can depend on adopted conventions, though, as we can see in Fig. 1.11(b), where the size of the circuit can be 2 (two concatenated $\sqrt{\text{NOT}}$s) and 1 (NOT gate).

The behavior of qubits in the above circuit's device—the Mach–Zehnder interferometer—teaches us that by a particular qubit arrangement in appropriate devices, we can overcome the detection problems we had with single beam splitters and polarizers, and that we can make use of the superposition of qubits in one step, allowing us to make use of many possible qubit states simultaneously. In the next section, we will see how we can use qubits for computation.

1.15 Qubit Computation and E-Business

Quantum computation suffer from the same disease that the first computers had. Recall Aiken's words, which we cited in the Introduction: "Only six electronic digital computers would be required to satisfy the computing needs of the entire United States." When the head of the computing center at Harvard uttered these words in 1947, there were simply no algorithms to formulate all of today's applications in the Boolean 0–1 language. Similarly, today, at the beginning of the quantum computing era, we have only one really convincing algorithm for classical applications: Peter Shor's (Sec. 3.3.3, p. 180) polynomial-time algorithm for prime factorization, whose known classical counterparts are of the exponential-time type [Shor, 1997]. There are also other algorithms such as Deutsch's (Sec. 3.3.1, p. 173), Deutsch-Jozsa (Sec. 3.3.2, p. 176), Bernstein-Vazirani (p. 179), Simon's (p. 180), and Grover's (p. 186). However, Shor's and other quantum algorithms reveal that we can use superposition in a way that gets around the statistics we have seen with a single beam splitter or polarizer.

In Sec. 1.10 (Fig. 1.9), we have seen that we can verify the "superposition" given by Eq. (1.18) so as to register photons with detectors D_0 and D_1. The statistics then give 50:50 odds of a photon being detected by one of the detectors. Which of the two detectors a photon will reach is unpredictable: the registration events are genuinely random. However, we can recombine the state vectors from Eq. (1.18) at another beam splitter, as shown in Sec. 1.14 in Eqs. (1.52) and (1.53), to obtain a registration with probability 1. Now the question is whether we can do such "single step recombinations" for nontrivial outcomes on the very same device, i.e., whether a Mach–Zehnder interferometer can mimic a "quantum transistor" and process arbitrarily many intermediate states. To achieve such a goal, we first make use of phase shifters set at chosen phase shifts as follows.

Let us set ϵ_0, ϵ_1, ϵ'_0, and ϵ'_1 so as to make phase shifts (with respect to the state of the incoming photon) ϕ_0, ϕ_1, ϕ'_0, and ϕ'_1, respectively. The operator describing the action of linear optical elements (beam splitters and phase shifters) in Fig. 1.10 (p. 29) is then

$$\begin{aligned}\hat{B} &= \begin{bmatrix} e^{i\phi'_0} & 0 \\ 0 & e^{i\phi'_1} \end{bmatrix} \frac{1}{\sqrt{2}} \begin{bmatrix} 1 & i \\ i & 1 \end{bmatrix} \begin{bmatrix} e^{i\phi_0} & 0 \\ 0 & e^{i\phi_1} \end{bmatrix} \frac{1}{\sqrt{2}} \begin{bmatrix} 1 & i \\ i & 1 \end{bmatrix} \\ &= e^{i(\phi_0+\phi_1\pi)/2} \begin{bmatrix} -e^{i\phi'_0}\sin\frac{\phi_1-\phi_0}{2} & e^{i\phi'_0}\cos\frac{\phi_1-\phi_0}{2} \\ e^{i\phi'_1}\cos\frac{\phi_1-\phi_0}{2} & e^{i\phi'_1}\sin\frac{\phi_1-\phi_0}{2} \end{bmatrix}. \end{aligned} \tag{1.57}$$

Setting the phase shifters ϵ'_0 and ϵ'_1 at $\phi'_0 = \phi'_1 = -(\phi_0 + \phi_1 + \pi)/2$ and denoting B for this special choice by R, we get a rotary action of the device on the input state vector $|0\rangle$:

$$\hat{R}\begin{bmatrix} 1 \\ 0 \end{bmatrix} = \begin{bmatrix} -\sin\frac{\phi_1-\phi_0}{2} & \cos\frac{\phi_1-\phi_0}{2} \\ \cos\frac{\phi_1-\phi_0}{2} & \sin\frac{\phi_1-\phi_0}{2} \end{bmatrix}\begin{bmatrix} 1 \\ 0 \end{bmatrix} = \begin{bmatrix} \sin\frac{\phi_0-\phi_1}{2} \\ \cos\frac{\phi_0-\phi_1}{2} \end{bmatrix}. \quad (1.58)$$

The circuit diagram corresponding to this equation is given in Fig. 1.12.

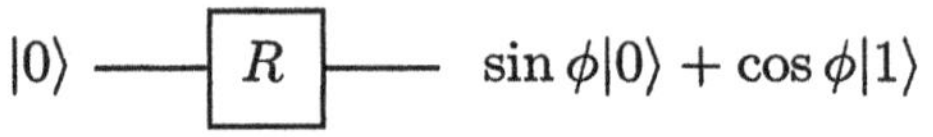

$$|0\rangle \; \text{---} \boxed{R} \text{---} \; \sin\phi|0\rangle + \cos\phi|1\rangle$$

Figure 1.12. One-qubit gate that turns photon qubits $|0\rangle$ into a superposition of $|0\rangle$ and $|1\rangle$ according to Eq. (1.58), where $\phi = (\phi_1 - \phi_0)/2$.

However, in the quantum computer literature, the so-called *Hadamard* and *phase gates*, respectively defined as

$$\hat{\mathrm{H}} = \frac{1}{\sqrt{2}}\begin{bmatrix} 1 & 1 \\ 1 & -1 \end{bmatrix}, \quad \text{and} \quad \hat{\Phi}_{[\phi_1][\phi_2]} = \frac{1}{\sqrt{2}}\begin{bmatrix} e^{i\phi_1} & 0 \\ 0 & e^{i\phi_2} \end{bmatrix}, (1.59)$$

are the gates one would use instead of R. A Hadamard gate can be made by means of a beam splitter and two phase shifters, the first at one of the entrances and the second at one of the exits:

$$\hat{\mathrm{H}} = \begin{bmatrix} 1 & 0 \\ 0 & e^{i\pi/2} \end{bmatrix}\frac{1}{\sqrt{2}}\begin{bmatrix} 1 & i \\ i & 1 \end{bmatrix}\begin{bmatrix} 1 & 0 \\ 0 & e^{i\pi/2} \end{bmatrix}. \quad (1.60)$$

The setup can also be reversed:

$$\frac{1}{\sqrt{2}}\begin{bmatrix} 1 & i \\ i & 1 \end{bmatrix} = \begin{bmatrix} 1 & 0 \\ 0 & e^{i\pi/2} \end{bmatrix}\hat{\mathrm{H}}\begin{bmatrix} 1 & 0 \\ 0 & e^{i\pi/2} \end{bmatrix} = \hat{\Phi}_{0\frac{\pi}{2}}\hat{\mathrm{H}}\hat{\Phi}_{0\frac{\pi}{2}}. \quad (1.61)$$

We can get our superposition of Eq. (1.58) with the help of these Hadamard gates, as shown in the circuit diagram of Fig. 1.13.

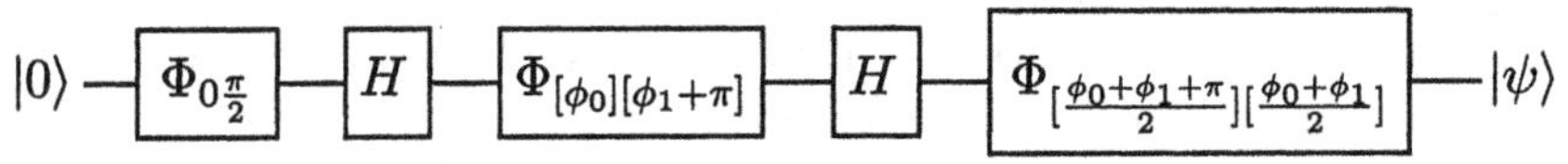

$$|0\rangle \text{---}\boxed{\Phi_{0\frac{\pi}{2}}}\text{---}\boxed{H}\text{---}\boxed{\Phi_{[\phi_0][\phi_1+\pi]}}\text{---}\boxed{H}\text{---}\boxed{\Phi_{[\frac{\phi_0+\phi_1+\pi}{2}][\frac{\phi_0+\phi_1}{2}]}}\text{---}|\psi\rangle$$

Figure 1.13. Quantum circuit diagram for turning qubit $|0\rangle$ into the superposition $[\sin\frac{\phi_0-\phi_1}{2} \; \cos\frac{\phi_0-\phi_1}{2}]^T$ given in Fig. 1.12.

The probability of detector D_1 registering a photon follows from Eqs. (1.41) and (1.58):

$$p_1 = \cos^2\frac{\phi_1 - \phi_0}{2} = \frac{1}{2}(1 + \cos\phi), \quad (1.62)$$

where $\phi = \phi_1 - \phi_0$. Note that the probability would stay the same if we took out the phase shifters ϵ_0' and ϵ_1'; therefore the result depends only on the phase difference $\phi_1 - \phi_0$.

Let us see how we can use the result to factor numbers in order to illustrate Shor's algorithm (short of entanglement and the corresponding speedup, which we are going to address later on), following [Summhammer, 1997].

We obtain the factors of a chosen number, say N, in a "physical" way using the setup shown in Fig. 1.10 (p. 29) and Eq. (1.62). Let us increase the phase shift ϕ in discrete steps $2\pi/n$ so as to have $\phi_k = 2\pi kN/n$, $k = 1, \ldots$ If we let n photons through the device: $k = 1, \ldots, n$, the sum of all individual probabilities that the detector D_1 would register a photon—given by Eq. (1.62)—will be

$$I_n = \sum_{k=1}^{n} p_1(k) = \frac{1}{2}\left[n + \sum_{k=1}^{n} \cos\left(\frac{2\pi kN}{n}\right)\right]. \tag{1.63}$$

If n were a factor of N we would have $p_1(k) = 1$ and $I_n = n$. If not, the cosines would roughly cancel each other and we would get $I_n \approx n/2$. Actually, we do not have to make all the measurements for the above sum because when n is a factor of N, we would ideally always get a click in D_1 and never in D_0, and when n is not a factor of N, we on average get half of the clicks in D_1 and half in D_0. Hence, as soon as we detect a series of clicks only in D_1, we simply check whether n is a factor of N.

The numbers we can factor in this way are not big, but the result is very instructive for understanding the problems we face with classical computers and the way we can solve them with quantum computers. For light with $\lambda = 500$ nm, if we use a *continuous wave (CW) laser* (for example, Nd:YAG) with which we can have a *coherence length*, Δl—the length over which the phase is fairly constant—of up to 300 km, the corresponding *coherence time* is $\Delta t = \Delta l/c$. The Heisenberg uncertainty relation for energy and time $\Delta E \Delta t \geq \hbar$ together with the Planck postulate $E = h\nu$ gives $\Delta\nu\Delta t \approx 1/4\pi$, where $\Delta\nu$ is called the *bandwidth.* From $c = \nu\lambda$ by differentiation we get $\Delta\lambda = -c\Delta\nu/\nu^2 = -\lambda^2\Delta\nu/c$, where $\Delta\lambda$ is called the *linewidth.* Dropping the minus sign, which shows only that the changes in $\Delta\nu$ and $\Delta\lambda$ are opposite, and using the previous relations, we get $\Delta l \approx \lambda^2/\Delta\lambda$. To keep the linewidth at $\Delta\lambda \approx 10^{-17}$ is feasible, since this linewidth corresponds to the coherence length $\Delta l \approx$ 25 km (cf. Sec. 2.7).

In our setup, at each phase step $\Delta\phi = 2\pi/n$, a photon is sent into the interferometer. The phase difference in $\Delta\phi$ in our interferometer proportional to $\Delta o/\lambda$, where Δo is the *optical path difference* [Born and

Wolf, 1997]. The Δo must be smaller than the coherence length and we can estimate that $n < \lambda/\Delta\lambda$.

Hence the biggest numbers we could factor are $N \approx 10^{10}$, and any PC can factor a number with 10 digits in a fraction of a second. However, the important property of this example of physical computing is that our "transistor"—Mach–Zehnder interferometer—is faster per computing unit (quantum gate) than the standard classical transistor for the same "clock" speed.

The longest factorization test, according to Eq. (1.63), will take time proportional to nN, because the maximum value of k is n. Since the largest n we have to check is $\sqrt{N}$, the maximum time would be proportional to $N^{3/2}$. However, as we stressed in the discussion following Eq. (1.63), a much smaller number of verifications, say 10, would also suffice. The required time is therefore a linear function of N.

A direct—and the most inefficient—algorithm for factoring a number would simply be $\sqrt{N}$ trial divisions.[8] Hence the number of checks that the most inefficient classical factoring algorithm has to carry out is smaller than the $10N$ we obtained for our "physical calculation" above. Still, given the same clock frequency, a classical computer calculation is slower per computing unit (gate). There are two reasons for this. First, we have to turn numbers into bits (and this is what makes the complexity of factoring numbers exponential—see Figs. 1.14 (a) and (b)), and then we have to carry out binary operations that correspond to division (which is one of the most complicated basic computer operations).

As opposed to a computer search–verify procedure, the photon search–verify Mach–Zehnder factorization procedure is instantaneous for each photon. The problem is that we cannot calculate much with only one Mach–Zehnder interferometer. We could parallelize the calculation by putting another Mach–Zehnder interferometer at each output of the first one, then putting another Mach–Zehnder interferometer at each output of the previous one, and so on [Summhammer, 1997]. However, that process would give rise to an exponentially growing number of elements, causing us to lose the advantage we gained. We will show how to get around this problem in Section 1.16. But before we dwell on the solution to the problem, we will say a bit more about another issue that requires an exponentially growing time for its solution and which is one of the major impetuses for quantum computer projects today.

As we mentioned, the two main quantum computation algorithms for classical applications are Shor's factoring algorithm and Grover's search

[8] We verify whether $xy = N$, starting with $x = 1$ and $y = N$ and ending with $x = y = \sqrt{N}$.

algorithm (Grover's algorithm can have quantum applications as well). These are so important not because they exemplify what a quantum computer could do but primarily because encryption and data search underlie e-business and e-communication today. Of all encryption on the Internet today 95% is based on the 512-bit RSA (Security, Inc., named after Rivest, Shamir, and Adleman [Rivest et al., 1978]) keys (containing 155 digits), i.e., on 155-digit composite numbers. Recently, several very efficient classical factoring algorithms have been developed. The fastest of these is the so-called *general number field sieve* or GNFS algorithm [Pomerance, 1996]. It enables a classical computer to factor a number N within a time frame proportional to the following complexity:

$$\exp\left(1.923(\log N)^{1/3}(\log\log N)^{2/3}\right). \tag{1.64}$$

In Fig. 1.14, the complexity of the GNFS algorithm (c) is compared to the complexity of trial division (a). We can see that a kind of "saturation" behavior is much more apparent for the complexity by the expression (1.64) than by $\sqrt{N}$. As a result, one can factor numbers within time frames that are many orders of magnitude smaller than the time frames needed to carry out our trial divisions. Fig. 1.14 compares numbers in the vicinity of 2^{512}, on which current RSA cryptographic keys are based.

In plots (b) and (d) of Fig. 1.14, functions (a) and (c), respectively, are shown as functions of the binary input size (number of bits, here from 500 to 512). The pattern we obtain in (b) is exponential, $2^{n/2}$, showing what is meant when one says that the factoring problem is an "exponential problem" or that it is a problem whose solution requires exponential time. When we do the same thing with the GNFS complexity, we obtain a *subexponential* [9] behavior:

$$\exp\left(2.45n^{1/3}\log n^{2/3}\right),$$

whose plot is shown in Fig. 1.14(d). So, for today's RSA keys, i.e., $n = 512$, we get 2×10^{19} for GNFS as opposed to 3.2×10^{77} for trial division. This reduction enabled the "cracking" of a challenge 512-bit RSA key in 1999 [Cavallar et al., 2000].

It is widely believed that a classical polynomial algorithm for computer factoring does not exist, although no one has proved this so far. Therefore the reaction to the cracking was, "Okay, let's use a 1024-bit

[9] Growing slower—as a function of n—than any exponential exponential function, but faster than any polynomial in n, for example, $\exp[n^\alpha]$, where $\alpha < 1$.

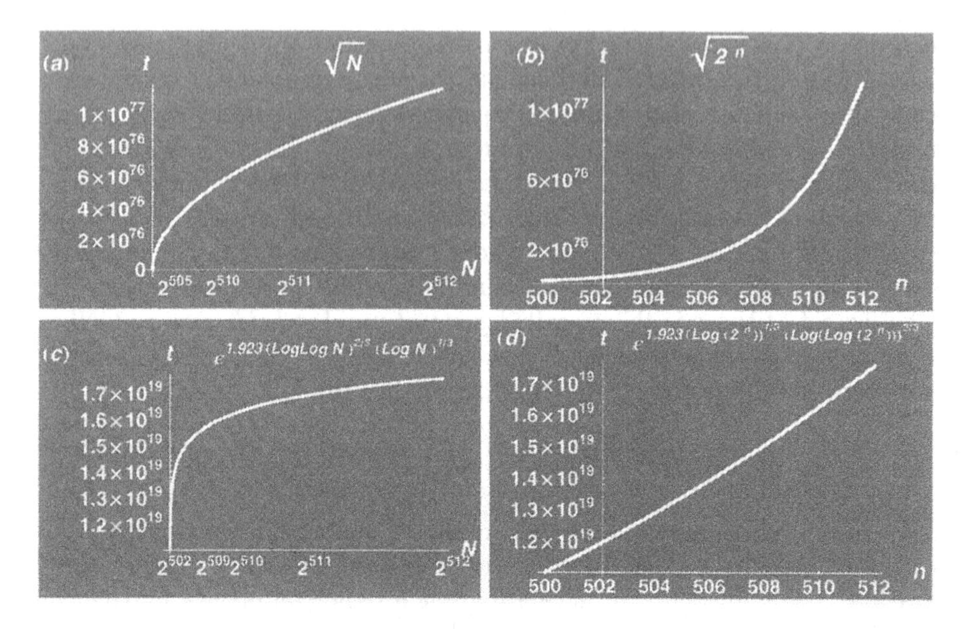

Figure 1.14. Complexities of factoring number N having n bits in the binary input size. (a) Trial division; (b) Trial division complexity expressed terms of n (the number of bits needed to write down N in a binary form); (c) Complexity of the GNFS algorithm ; (d) GNFS algorithm complexity expressed in terms of n.

RSA." The nonclassical Shor's algorithm, which is polynomial and which surpasses the approach of the GNFS algorithm, could crack any such key on would-be quantum computers in seconds. However, as far as Internet security is concerned, efforts to make quantum computers work are not motivated by a desire to make the Internet unsafe but rather to provide it with much safer technology—*quantum cryptography*—which is based on quantum communication and which apparently can work more reliably when coupled to a quantum computer. Any night someone could come forward with an ingenious classical polynomial algorithm and make the Internet instantly vulnerable the very next morning. Recall that in 1976, in Martin Gardner's *Scientific American* column, the 129-digit RSA key was estimated to be safe for 40 quadrillion years [Pomerance, 1996]), but by December 3, 2003, the 576-bit RSA key (174 digits) had already been cracked [Weisstein, 2003].

1.16 Numbers and Bits

We emphasized that the physical processing of numbers per quantum gate (processing unit, Mach–Zehnder interferometer) takes much less time than the binary processing of numbers per classical (logic) gate.

The main difference is that quantum qubit encoding is based on gates that can process and produce arbitrarily many values of a single qubit and even arbitrarily many qubits, while classical bit encoding is based on two-valued, binary gates that can process only two bits at a time. To understand the difference, we will show some details of digital bit processing. In doing so, we want to pave the road for a comparison between the classical approach to gates reviewed here and the quantum proposals reviewed in Chapter 2.

The binary representation of a decimal number N is given by a binary digit string

$$N_2 = \alpha_{n-1}\alpha_{n-2}\dots\alpha_1\alpha_0, \tag{1.65}$$

where α_i, $i = 0, \dots, n-1$ are determined from the following equation:

$$N_2 = \alpha_{n-1}2^{n-1} + \alpha_{n-2}2^{n-2} + \dots + \alpha_1 2^1 + \alpha_0 2^0 = \sum_{i=0}^{n-1} \alpha_i 2^i. \tag{1.66}$$

So, if we want to obtain a representation of a number, say 36, we can consecutively divide the number by 2 as follows:

$$\begin{array}{r|lll}
36 & \text{remainder } = 0 \Rightarrow & \alpha_0 = 0, & \text{LSB, least significant bit} \\
18 & \text{remainder } = 0 \Rightarrow & \alpha_1 = 0, & \\
9 & \text{remainder } = 1 \Rightarrow & \alpha_2 = 1, & \\
4 & \text{remainder } = 0 \Rightarrow & \alpha_3 = 0, & \\
2 & \text{remainder } = 0 \Rightarrow & \alpha_4 = 0, & \\
1 & \Rightarrow & \alpha_5 = 1, & \text{MSB, most significant bit.}
\end{array} \tag{1.67}$$

Hence, the binary representation of 36 is 100100. In the other direction, the decimal representation of 100100 is $2^2 + 2^5 = 36$.

To handle numbers by means of gates, i.e., transistors, we have to manipulate strings of bits representing those numbers as well as the outcomes of arithmetic operations carried out on them. The proper combination of gates is required to manipulate these strings and to carry out addition. (Other operations can be reduced to addition.) When we try to add single bits, $0 + 0 = 0$, $0 + 1 = 1 + 0 = 1$, we realize that $1 + 1$ requires a 2-bit string: $1 + 1 = 10$. We get the string using the so-called *half adder* shown in Fig. 1.15(a), where the sum $\mathbf{S}=\mathbf{A}\oplus\mathbf{B}$ in the last case is 0 and the *carry* $\mathbf{C}_{\text{out}}=\mathbf{AB}$ is 1.[10] For bigger numbers, we need to reuse the carry. The *full adder* shown in Fig. 1.15(b) serves this purpose.

[10]So, we have the following strings: $\{\mathbf{C}_{\text{out}}\}\{\mathbf{S}\} = 00$ (for $0+0$), 01 (for $0+1=1+0$), and 10 (for $1+1$).

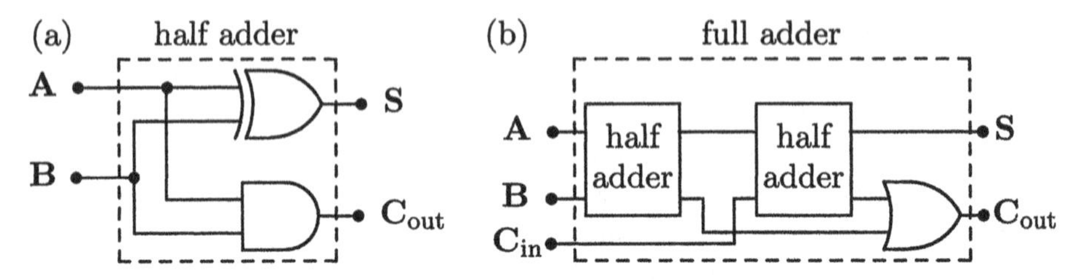

Figure 1.15. (a) $\mathbf{S=A\oplus B}$, $\mathbf{C_{out}=AB}$; (b) $\mathbf{S=(A\oplus B)\oplus C_{in}}$, $\mathbf{C_{out}=(A\oplus B)C_{in}+AB}$.

Full adders are combined in blocks so as to form *binary adders*. The binary adder shown in Fig. 1.16 is an 8-bit (1-byte) device, and we see that it is capable of dealing with any number whose sum does not exceed 255. Otherwise we get an *overflow* ($\mathbf{C_{fo}}$=1 in Fig. 1.16).

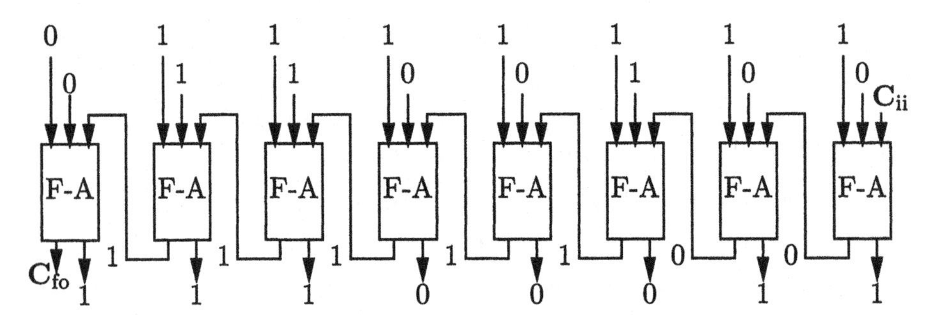

Figure 1.16. Eight-bit binary adder. F-A are full adders (Fig. 1.15(b)). $\mathbf{C_{ii}}$ and $\mathbf{C_{fo}}$ are initial input and final output carry bits, respectively. The addition 127+100=227 is shown in the binary representation: 01111111+01100100=11100011; $\mathbf{C_{ii}=C_{fo}=0}$.

Figs. 1.15 and 1.16 illustrate why binary processing can take so much more time than "physical" processing. A transistor is a switch—see Fig. 1.1 (p. 10)—with a delay between an input voltage change and the response on its output. This time is called *gate delay*. Delay times in transistors can be brought down to less than 1 ns, but let us use 1 ns as an illustration. This time may be slowed down to about 2 ns for some gates (typically, NAND is one of the fastest and XOR is the slowest). Since a full adder is a cascade of two half adders, the former might yield a result within 4 to 8 ns, since the time varies with inputs (1s or 0s). This time increases linearly with the bit length because each full adder within binary adders has to wait for the carry from a previous stage (see Fig. 1.16) to output a steady-state result. Thus, solutions of problems for which complexity grow exponentially with time require both an exponential reduction of gate delays and an exponential increase of the number of transistors. We have already seen that further reductions in the size of transistors, and thereby increases in their numbers within processors, will soon reach their limits. Attempts to further reduce gate delays will also hit its physical limits.

On the other hand, the physical calculation we considered in Sec. 1.15 is based on a single qubit and a single gate. Due to quantum superposition, we were able to put a number composed of 10 digits, i.e., 34 bits, through a single gate in a single step. Classically this process corresponds to 34 gates, or transistors wired in parallel, each of which is slower than its quantum counterpart. Ideally, quantum superposition would allow an unlimited exponential speedup. However, as we have already shown in Sec. 1.15 (p. 31), the main problem of realistic quantum physical computing is to scale it up. We will address the general problem of the scalability of would-be quantum computers in Chapter 2. In the next section, we will see how one can improve quantum physical calculation itself.

1.17 Entangled Qubits

In Sec. 1.15 we emphasized that a single quantum gate (exemplified by the Mach–Zehnder interferometer) processing a single qubit (photon) has a limited output size. The size can be increased by putting other Mach–Zehnder interferometers at outputs of the previous ones, but that would amount to exponentially increasing the number of elements. Another way is to increase the number of qubits that we process simultaneously. In our case, this means that at least two photons arrive together at a beam splitter.

We can manipulate incoming photons using their mutual phase shifts as well as their polarization. In Sec. 1.15, we used phase shifts. In this section, we are going to combine phase shifts at a beam splitter with polarization so as to better understand a general formalism we are going to introduce in Sec. 1.18. Unpolarized photons arrive simultaneously at a beam splitter, as shown in Fig. 1.17 [Pavičić, 1994]. Realistic experiments feeding a beam splitter with unpolarized photons are feasible and have, in effect, been carried out many times as an inherent part of many four-photon experiments, for example, in [Bouwmeester et al., 1997].

To obtain a general result, we start with polarized photons, from which we obtain in the end the results for unpolarized photons. Polarized photons arriving at a beam splitter are described by the product

$$\begin{aligned} |\Psi\rangle &= (\cos\theta_{1'}|0\rangle_1 + \sin\theta_{1'}|1\rangle_1)\otimes(\cos\theta_{2'}|0\rangle_2 + \sin\theta_{2'}|1\rangle_2) \\ &= \cos\theta_{1'}\cos\theta_{2'}|0\rangle_1\otimes|0\rangle_2 + \cos\theta_{1'}\sin\theta_{2'}|0\rangle_1\otimes|1\rangle_2 \\ &\quad + \sin\theta_{1'}\cos\theta_{2'}|1\rangle_1\otimes|0\rangle_2 + \sin\theta_{1'}\sin\theta_{2'}|1\rangle_1\otimes|1\rangle_2\,, \end{aligned} \tag{1.68}$$

where $\theta_{1'}$,$\theta_{2'}$ are the angles along which incident photons are polarized with respect to a fixed direction ($|0\rangle$ and $|1\rangle$ denote the mutually orthogonal photon states).

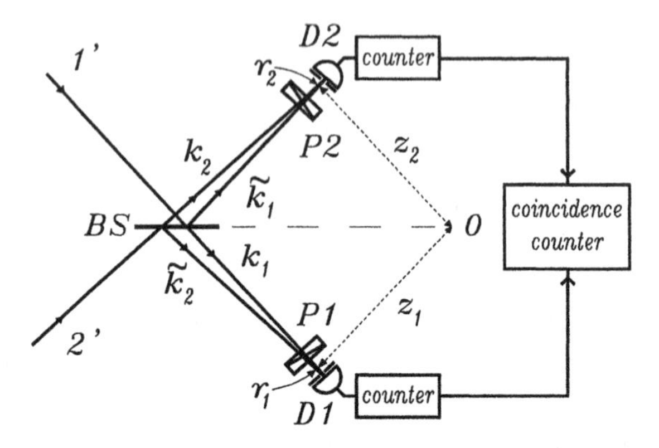

Figure 1.17. Two unpolarized incoming photons maximally entangled in polarization upon leaving the beam splitter, BS. Figure taken from [Pavičić, 1994].

$|\Psi\rangle_1 \otimes |\Phi\rangle_2$ that appears in Eq. (1.68) is called the *tensor product* of vectors $|\Psi\rangle_1$ and $|\Phi\rangle_2$ describing two different quantum systems and therefore belonging to two different Hilbert spaces, say $\mathcal{H}_1$ and $\mathcal{H}_2$. In the rest of this book, we shall often drop the sign $\otimes$, in particular when writing a product of *pure states* (which cannot be written as a nontrivial linear combination of two independent states), for example, $|0\rangle_1|0\rangle_2$, etc. in Eq. (1.68). $|0\rangle_1$ denotes the state of an upper incoming photon polarized in the x-direction (we might call it a horizontal or vertical polarization according to our choice). Hence, $|0\rangle$ and $|1\rangle$ do not mean 0 and 1 number states in the Fock space as in quantum optics, but rather both $|0\rangle$ and $|1\rangle$ mean a one-photon state.

For those familiar with the Fock space notation, we just mention here that in it we would write $|1_x\rangle$ and $|1_y\rangle$ for $|0\rangle$ and $|1\rangle$, respectively. If the beam splitter were removed, it would cause a "click" at detector D1 and no "click" at detector $\mathrm{D1}^\perp$, provided the birefringent polarizer P1 is oriented along x. Here $\mathrm{D1}^\perp$ means a detector counting photons coming out at the *other exit* $\mathrm{P}^\perp$ (perpendicular polarization; not shown in Fig. 1.17) of the birefringent prism P1.

To describe the joint actions of polarizers, beam splitter, and detectors, we shall use the so-called *annihilation* and *creation* operators (taken from the so-called second quantization formalism; however, the reader is not expected to be familiar with the formalism). The operators act on our states as follows: $\hat{a}_{1x}|0\rangle_1 = |\varnothing\rangle_1$, where $|\varnothing\rangle_1$ means a state without a photon. This is an *empty state* or a *vacuum state* in Fock notation. It describes the detection of a photon and formalizes the calculation of the probabilities of detecting photons ($\langle\varnothing|\varnothing\rangle_1 = 1$). The reader is not expected to be familiar with Fock notation. Instead, we should consider the following equations as self-explanatory definitions of our annihilation (a) and its conjugate ($a^\dagger$) creation operators: $\hat{a}_{1x}|0\rangle_1 = |\varnothing\rangle_1$, $\hat{a}_{1x}|\varnothing\rangle_1 = 0$, $\hat{a}_{1x}|1\rangle_1 = 0$, $\hat{a}_{1y}|0\rangle_1 = 0$, $\hat{a}_{1y}|\varnothing\rangle_1 = 0$, $\hat{a}_{1y}|1\rangle_1 = |\varnothing\rangle_1$, $\hat{a}_{2x}|0\rangle_2 = |\varnothing\rangle_2$, $\hat{a}_{2x}|0\rangle_1 = 0$, etc., ${}_1\langle 0|\hat{a}^\dagger_{1x} = {}_1\langle\varnothing|$, etc.

When we look at photons leaving the beam splitter, there are two possibilities: photons are detected either at the opposite or at the same sides of the beam splitter. The outgoing electric-field operator (a simple plane-wave description with "annihilation operator designation of virtual photon paths") for the first possibility for the upper outgoing path reads

$$\begin{aligned}\hat{E}_1 &= (\hat{a}_{1x}t_x\cos\theta_1 + \hat{a}_{1y}t_y\sin\theta_1)\,e^{i\mathbf{k}_1\cdot\mathbf{r}_1 - i\omega_1(t-\tau_1)} \\ &\quad + i\,(\hat{a}_{2x}r_x\cos\theta_1 + \hat{a}_{2y}r_y\sin\theta_1)\,e^{i\tilde{\mathbf{k}}_2\cdot\mathbf{r}_1 - i\omega_2(t-\tau_2)}\,, \qquad (1.69)\end{aligned}$$

where θ_1 and θ_2 from Eq. (1.70) are the angles along which the polarizers in front of detectors D1 and D2 in Fig. 1.17 are oriented, τ_j is time delay after which the photon reaches detector D$_j$; ω_j is the frequency of photon j, $j = 1, 2$; and t and r are *transmission* and reflection coefficients, respectively at the beam splitter BS. Eq. (1.70) describes the superposition of the path being reflected from the beam splitter (incoming from above) with the path being transmitted through the beam splitter (incoming from below). The action of the beam splitter, as, for example, in Eq. (1.54), has been taken into account, only here we have a not necessarily symmetrical beam splitter with $t = |\sqrt{T}|$ and $r = |\sqrt{R}|$, where T and R denote transmittance and reflectance, respectively. E_2 for the upper path is obtained analogously.

The outgoing electric-field operator for the second photon emerging from the beam splitter at the same side as the first one reads (for the upper path)

$$\begin{aligned}\hat{E}'_2 &= (\hat{a}_{2x}t_x\cos\theta_2 + \hat{a}_{2y}t_y\sin\theta_2)\,e^{i\mathbf{k}'_2\cdot\mathbf{r}'_2 - i\omega_2(t-\tau_2)} \\ &\quad + i\,(\hat{a}_{1x}r_x\cos\theta_2 + \hat{a}_{1y}r_y\sin\theta_2)\,e^{i\tilde{\mathbf{k}}'_1\cdot\mathbf{r}'_2 - i\omega_1(t-\tau_1)}\,. \qquad (1.70)\end{aligned}$$

E'_1 for the lower path is obtained analogously.

Choosing $t_x = t_y = r_x = r_y = 2^{-1/2}$ and $z_1 = z_2$, we obtain the probability of detecting one photon by detector D1 and the other by D2:

$$\begin{aligned}P(\theta_{1'},\theta_{2'},\theta_1,\theta_2) &= \langle\Psi|\hat{E}^\dagger_2\hat{E}^\dagger_1\hat{E}_1\hat{E}_2|\Psi\rangle \\ &= \frac{1}{4}\sin^2(\theta_{1'} - \theta_{2'})\sin^2(\theta_1 - \theta_2)\,. \qquad (1.71)\end{aligned}$$

We see that the probability factors left–right and not up–down, as one would be tempted to conjecture from the initial up-down independence expressed by the product of the "upper" and "lower" function in Eq. (1.68).

The probability of both photons being detected by D2 is

$$P(\theta_{1'}, \theta_{2'}, \theta_1 \times \theta_2) = \frac{1}{2}\langle\Psi|\hat{E}_2^\dagger \hat{E}_2'^\dagger \hat{E}_2' \hat{E}_2|\Psi\rangle$$
$$= \frac{1}{8}[\cos(\theta_{1'}-\theta_2)\cos(\theta_{2'}-\theta_1)+\cos(\theta_{1'}-\theta_1)\cos(\theta_{2'}-\theta_2)]^2. \quad (1.72)$$

For unpolarized photons, the density matrix is proportional to the unit matrix, and this means that we only need the products

$$|0\rangle_1\,|0\rangle_2,\ |0\rangle_1\,|1\rangle_2,\ |1\rangle_1\,|0\rangle_2,\ \text{and}\ |1\rangle_1\,|1\rangle_2, \quad (1.73)$$

to form partial probabilities, which then sum up to the total correlation probability. Hence, $\hat{E}_1, \hat{E}_2$ as given by Eq. (1.69) should be applied to $|0\rangle_1|0\rangle_2$, $|0\rangle_1|1\rangle_2$, $|1\rangle_1|0\rangle_2$, and $|1\rangle_1|1\rangle_2$ so as to give four probabilities, which then sum up to the following correlation probability:

$$P(\infty, \infty, \theta_1, \theta_2) = \frac{1}{8}\sin^2(\theta_2 - \theta_1)\,. \quad (1.74)$$

The overall probability of their appearance on the same side of the beam splitter is

$$P(\infty, \infty, \theta_1 \times \theta_2) = \frac{1}{8}\big[1 + \cos^2(\theta_1 - \theta_2)\big]. \quad (1.75)$$

This means that the photons appear on the same side with a greater probability than on opposite sides (see Fig. 1.18), as follows from Eqs. (1.75) and (1.71):

$$P(\infty,\infty,\theta_1,\theta_2)+P(\infty,\infty,\theta_1^\perp,\theta_2)+P(\infty,\infty,\theta_1,\theta_2^\perp)+P(\infty,\infty,\theta_1^\perp,\theta_2^\perp)$$
$$= P(\infty,\infty,\infty,\infty) = \frac{2}{8}[\cos^2(\theta_1-\theta_2)+\sin^2(\theta_1-\theta_2)] = \frac{1}{4},$$
$$P(\infty,\infty,\theta_1\times\theta_2)+P(\infty,\infty,\theta_1^\perp\times\theta_2)$$
$$+P(\infty,\infty,\theta_1\times\theta_2^\perp)+P(\infty,\infty,\theta_1^\perp\times\theta_2^\perp)$$
$$= P(\infty,\infty,\infty\times\infty) = \frac{2}{8}[1+\cos^2(\theta_1-\theta_2)+1+\ sin^2(\theta_1-\theta_2)] = \frac{3}{4}. \quad (1.76)$$

It follows from Eq. (1.74) that the photons that arrive unpolarized at the beam splitter emerge from it (anti)correlated in polarization whenever they appear at the opposite sides of the beam splitter. We say that they are *entangled.* They are described by the following state, which we

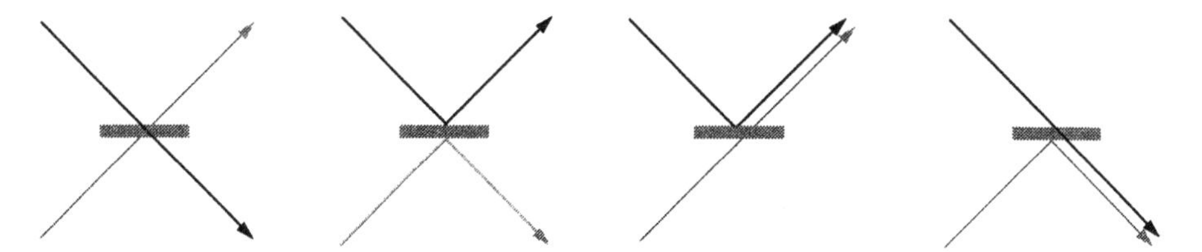

Figure 1.18. The first two and the second two possible photon paths do not contribute to equal probabilities of detection behind the beam splitter. The first probability is 0.25 and the second 0.75. See text.

will call the *singlet-like state*[11]:

$$|\Psi_s\rangle = \frac{1}{\sqrt{2}}(|0\rangle_1|1\rangle_2 - |1\rangle_1\,|0\rangle_2)\,. \tag{1.77}$$

We obtain this state by applying E_1 [Eq. (1.69)] and E_2 to the states given by Eq. (1.73). The probability of this state is then equal to probability (1.74) multiplied by 4 (for perpendicularly oriented polarizers). It is instructive to verify this outcome by means of the rules of thumb for the annihilation operator we introduced above.

We describe the actions of polarizer P1 (tilted by an angle θ_1 with respect to a chosen direction) and detector D1, and of polarizer P2 (tilted by an angle θ_2) and detector D2 by the operators $\mathcal{D}_1$ and $\mathcal{D}_2$, respectively:

$$\mathcal{D}_1 = a_{01}\cos\theta_1 + a_{11}\sin\theta_1, \qquad \mathcal{D}_2 = a_{02}\cos\theta_2 + a_{12}\sin\theta_2, \tag{1.78}$$

where we drop the "hats" of the annihilation operators a to ease the notation; a_{01} describes the action of P1 (which lets vertically (0) polarized photons through) on photon 1, a_{12} describes the action of P2 (which lets horizontally (1) polarized photons through) on photon 2, etc. Since

$$\mathcal{D}_1\mathcal{D}_2|\Psi_s\rangle = \frac{1}{\sqrt{2}}(\cos\theta_1\sin\theta_2 - \sin\theta_1\cos\theta_2)|\varnothing\rangle_1|\varnothing\rangle_2, \tag{1.79}$$

we get the probability of registering clicks in D1 and D2 as

$$\langle\Psi_s|\mathcal{D}_2^\dagger\mathcal{D}_1^\dagger\mathcal{D}_1\mathcal{D}_2|\Psi_s\rangle = \frac{1}{2}\sin^2(\theta_1 - \theta_2), \tag{1.80}$$

which is nothing but the probability (1.74) multiplied by 4 (if we included $\mathrm{P1}^\perp$ and $\mathrm{P2}^\perp$ as well, we would get exactly (1.74)).

[11]One of the so-called *Bell states.* See Eq. (3.40), p. 154.

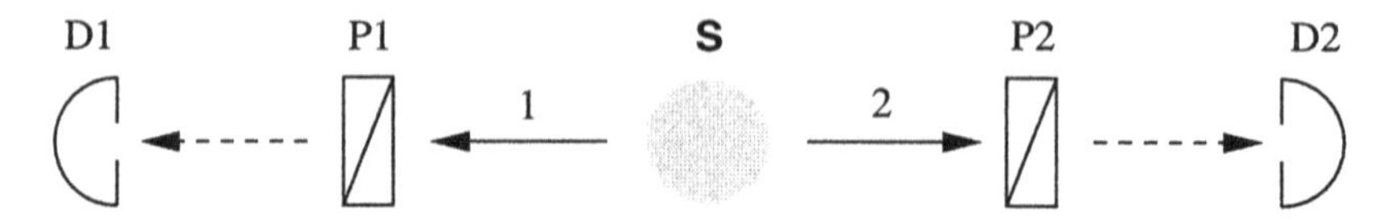

Figure 1.19. Entangled photons cannot be separated. Whatever their source—atom, beam splitter, or nonlinear crystal of type II (see Sec. 2.7, (p. 125)—and whatever the time window within which we randomly rotate the polarizers P1 and P2, for a singlet-like state (Eq. (1.77)), the clicks of detectors D1 and D2 will always show the probability correlation $\frac{1}{2}\sin^2(\theta_1 - \theta_2)$, as predicted by Eq. (1.74) (multiplied by 4).

We often call such an entangled pair of qubits an EPR (Einstein-Podolsky-Rosen) pair. The state (1.77) corresponds to probability (1.74) multiplied by 4 to match the selection carried out by disregarding the photons that appear from the same sides of the beam splitter and keeping only those pairs that emerge from its opposite sides.

This is the main clue for understanding entangled states. We engineer the entangled states by making a selection from all possible qubit states. Let us assume that the incoming photons are mutually perpendicularly polarized, for example, $\theta_{1'} = 0$ and $\theta_{2'} = \pi/2$ in Eq. (1.68). Then the state describing the paths of two photons (anti)correlated in polarization appearing at the opposite sides of the beam splitter together with those (anti)correlated in polarization appearing at the same sides of the beam splitter is given as

$$|\Psi\rangle = \frac{1}{2}(|0\rangle_1|1\rangle_2 - |1\rangle_1\,|0\rangle_2 + i|0\rangle_1|1\rangle_1 + i|0\rangle_2\,|1\rangle_2)\,. \tag{1.81}$$

The function can be verified by detectors D1 and D2 (D1$^\perp$, D2$^\perp$ are implicitly assumed). Its $|\Psi_s\rangle$ component given by Eq. (1.77), is to be verified by simultaneous detections at D1 and D2, while $|0\rangle_1|1\rangle_1$ ($|0\rangle_2|1\rangle_2$) has to be verified by a double detection at D1 (D2) (hard but feasible, see p. 157). The function (1.81) can be written as a product of two states:

$$|\Psi\rangle = \frac{1}{\sqrt{2}}(|0\rangle_1 + i|0\rangle_2) \otimes \frac{1}{\sqrt{2}}(i|1\rangle_1 + |1\rangle_2)\,. \tag{1.82}$$

However, its $|\Psi_s\rangle$ component, which corresponds to a selection from a complete set of measurements and data, cannot be written as a product. To prove this, let us assume it can:

$$\begin{aligned}|\Psi_s\rangle &= |0\rangle_1|1\rangle_2 - |1\rangle_1\,|0\rangle_2 = (\alpha_1|0\rangle_1 + \alpha_2|0\rangle_2)) \otimes (\beta_1|1\rangle_1 + \beta_2|1\rangle_2)\\ &= \alpha_1\beta_1|0\rangle_1|1\rangle_1 + \alpha_1\beta_2|0\rangle_1\,|1\rangle_2 + \alpha_2\beta_1|1\rangle_1|0\rangle_2 + \alpha_2\beta_2|0\rangle_2\,|1\rangle_2.\end{aligned}$$

We get $\alpha_1\beta_2 = -\alpha_2\beta_1 = 1$ and $\alpha_1\beta_1 = \alpha_2\beta_2 = 0$, which is a contradiction.

The above serves as a general definition of entanglement.

An *entangled state* is a correlated state that is not a product state, i.e., it cannot be separated (see Figs. 1.19, p. 44 and 1.24, p. 56).

We will see in Sec. 1.20 (p. 56) that we can describe entangled qubits by means of quantum circuits that would require that two or more lines—as shown in Fig. 1.24, p. 57—correspond to one inseparable, entangled state. Often we obtain entanglement not just by letting qubits through a gate (as we did with the beam splitter) but also as a result of changing a collective state of many qubits together so that we may have entangled qubits per line (see Fig. 1.31, p. 80. Nevertheless, to achieve an unambiguous representation, we shall introduce a more formal description first of single and then of many qubits so as to connect the formalism of Secs. 1.15 (p. 31) and 1.17 (p. 31) and enable a description of entangled qubits.

1.18 General Single Qubit Formalism

Our aim in this section is to arrive at the general qubit formalism starting from physically feasible optical models. To describe the behavior of two photons at a beam splitter and their subsequent passage through polarizers, we used Eqs. (1.69) and (1.70), which implicitly referred to the standard description of a nonsymmetrical beam splitter and the phase shifts that the photon fields can gain from it. In general, such a description is given by

$$\begin{bmatrix} \hat{a}_{1out} \\ \hat{a}_{2out} \end{bmatrix} = B \begin{bmatrix} \hat{a}_{1in} \\ \hat{a}_{2in} \end{bmatrix} = \begin{bmatrix} b_{11}e^{i\phi_{11}} & b_{12}e^{i\phi_{12}} \\ b_{21}e^{i\phi_{21}} & b_{22}e^{i\phi_{22}} \end{bmatrix} \begin{bmatrix} \hat{a}_{1in} \\ \hat{a}_{2in} \end{bmatrix}. \tag{1.83}$$

We do not include polarization, since it can be introduced afterwards and does not add to our point.

It can be shown that by invoking the conservation of energy of photons at the beam splitter, we easily get [Ou and Mandel, 1989; Campos et al., 1989]

$$B = e^{i\phi_0} \begin{bmatrix} \cos\theta\, e^{i\phi_t} & \sin\theta\, e^{i\phi_r} \\ -\sin\theta\, e^{-i\phi_r} & \cos\theta\, e^{-i\phi_t} \end{bmatrix}, \tag{1.84}$$

where

$$b_{11} = b_{22} = t = \cos^2\theta\,, \qquad b_{11} = b_{22} = t = \cos^2\theta\,, \tag{1.85}$$

$$\phi_t = \frac{1}{2}(\phi_{11} - \phi_{22})\,, \quad \phi_r = \frac{1}{2}(\phi_{12} - \phi_{22} \mp \pi)\,, \quad \phi_0 = \frac{1}{2}(\phi_{11} + \phi_{22})\,.$$

The determinant of B is

$$\det(B) = e^{2i\phi_0}, \tag{1.86}$$

and from Eq. (1.38) it follows that B is a unitary operator.

Hence, all the choices we can make for a photon gate (beam splitter) are equivalent. Thus we get

$$\begin{bmatrix} \cos\theta & i\sin\theta \\ i\sin\theta & \cos\theta \end{bmatrix}, \quad \begin{bmatrix} \cos\theta & \sin\theta \\ -\sin\theta & \cos\theta \end{bmatrix}, \tag{1.87}$$

for $\phi_0 = 0$, $\phi_t = 0$, and $\phi_r = \pi/2$, and for $\phi_0 = \phi_t = \phi_r = 0$, respectively. We also obtain the unit and Pauli matrices:

$$\mathbb{1} = \begin{bmatrix} 1 & 0 \\ 0 & 1 \end{bmatrix}, \quad \hat{\sigma}_x = \begin{bmatrix} 0 & 1 \\ 1 & 0 \end{bmatrix} = \mathrm{NOT} = \mathrm{X},$$
$$\hat{\sigma}_y = \begin{bmatrix} 0 & -i \\ i & 0 \end{bmatrix} = -i\mathrm{Y}, \quad \hat{\sigma}_z = \begin{bmatrix} 1 & 0 \\ 0 & -1 \end{bmatrix} = \mathrm{Z}, \tag{1.88}$$

for $\theta = 0$ and $\phi_0 = \phi_t = 0$, $\theta = \pi/2$, $\phi_0 = \pi/2$, and $\phi_r = -\pi/2$, $\theta = \pi/2$, $\phi_0 = 0$, and $\phi_r = -\pi/2$, and $\theta = 0$, $\phi_0 = \pi/2$, and $\phi_t = -\pi/2$, respectively, where X, Y, and Z are symbols more often used in quantum information theory than $\hat{\sigma}_x$, $\hat{\sigma}_y$, and $\hat{\sigma}_z$ [Cf. Eqs. (1.152–1.154)].

Moreover, it turns out that the unit and Pauli matrices can be used to express any 2×2 matrix α:

$$\alpha = \begin{bmatrix} \alpha_{11} & \alpha_{12} \\ \alpha_{21} & \alpha_{22} \end{bmatrix} = \sum_{i=1}^{3} c_i\hat{\sigma}_i + c_4\mathbb{1}, \tag{1.89}$$

Eq. (1.89) is a system of four equations with four unknowns, with the determinant of the coefficients being equal to $4i \neq 0$, which means that a nontrivial solution always exists. Hence, $\hat{\sigma}_i$ and $\mathbb{1}$ span the whole 2-dimensional space of 2×2 matrices.

It is well known that the Pauli operators correspond to the spin operators of $\frac{1}{2}\hbar$ in the following way:

$$\hat{S}_x = \frac{1}{2}\hbar\hat{\sigma}_x, \quad \hat{S}_y = \frac{1}{2}\hbar\hat{\sigma}_y, \quad \hat{S}_z = \frac{1}{2}\hbar\hat{\sigma}_z. \tag{1.90}$$

Both experiments and the so-called *linearization of the wave equation* (which dispenses with the relativistic origin of the spin) confirm that spin is an angular momentum, and it has to have a formal description analogous to a description of an orbital angular momentum in a 3-dimensional space up to the *gyromagnetic factor* (cf. p. 90). This factor makes the spin twice as large as it would be if it were a proper angular momentum of the orbital motion, and it is obtainable non-relativistically by means of the linearization of the wave equation [Greiner, 1989]. For the details, see Sec. 2.2 (p. 88). Hence, the spin operators and Pauli matrices are components of a 3-dimensional vector operator

$$\hat{\mathbf{S}} = \{\hat{S}_x, \hat{S}_y, \hat{S}_z\} = \frac{1}{2}\hbar\hat{\boldsymbol{\sigma}} = \frac{1}{2}\hbar\{\hat{\sigma}_x, \hat{\sigma}_y, \hat{\sigma}_z\}, \tag{1.91}$$

and $\hat{\boldsymbol{\sigma}}$ is a 3-dimensional vector whose components are 2×2 matrices.

Eq. (1.89), together with the physical meaning of the Pauli matrices given by Eqs. (1.90) and (1.91), enables us to obtain the following general expression of the *density matrix* for an arbitrary qubit:

$$\hat{\rho} = \frac{1}{2}(\mathbb{1} + \mathbf{r}\cdot\hat{\boldsymbol{\sigma}}), \tag{1.92}$$

where 1/2 was chosen to meet the requirement $\mathrm{Tr}(\rho)=1$ and $\mathbf{r}$ is a 3-dimensional vector $\mathbf{r} = \{x, y, z\}$, which assigns a 2×2 matrix

$$\mathbf{r}\cdot\hat{\boldsymbol{\sigma}} = r_x\sigma_x + r_y\sigma_y + r_z\sigma_z = \begin{bmatrix} z & x - iy \\ x + iy & -z \end{bmatrix} \tag{1.93}$$

to every point of our experimental 3-dimensional space.

The idea behind this particular form of $\mathbf{r}\cdot\hat{\boldsymbol{\sigma}}$ is to get a rotation in SO(3), and this rotation requires a transformation that will leave $\mathbf{r}^2$ invariant. The matrix $\mathbf{r}\cdot\hat{\boldsymbol{\sigma}}$ is therefore constructed so as to have its determinant proportional to $\mathbf{r}^2$. The transformation is then given by Eq. (1.95) below.

From the literature, one is sometimes tempted to conclude that, given a 3-dimensional vector $\mathbf{r}$, one can uniquely determine ρ by means of Eq. (1.92) and therefore the state of the corresponding qubit. This is, however, not so. To see why, we have to establish a morphism between the *special 2-dimensional unitary group* (2-dimensional unimodular unitary matrices), SU(2), and the *special orthogonal 3-dimensional rotation group*, SO(3) [Hammermesh, 1962].

Consider δ (cf. Eq. (1.89)) which is unitary ($\delta^{-1} = \delta^\dagger$) and unimodular ($\det\delta = 1$):

$$\delta = \begin{bmatrix} \delta_{11} & \delta_{12} \\ -\delta_{12}^* & \delta_{11}^* \end{bmatrix}, \qquad \text{where } \delta_{11}\delta_{11}^* + \delta_{12}\delta_{12}^* = 1. \tag{1.94}$$

Let us use this unitary matrix to define the following transformation of $\mathbf{r}\cdot\hat{\boldsymbol{\sigma}}$ from Eq. (1.93):

$$\mathbf{r}'\cdot\hat{\boldsymbol{\sigma}} = \delta\mathbf{r}\cdot\hat{\boldsymbol{\sigma}}\delta^\dagger. \tag{1.95}$$

This defines a transformation

$$\mathbf{r} \to \mathbf{r}' = R(\delta)\mathbf{r}. \tag{1.96}$$

Checking the determinant of this transformation,

$$\det \mathbf{r}' \cdot \hat{\boldsymbol{\sigma}} = -\det \mathbf{r}'^2 = \det \delta \mathbf{r} \cdot \hat{\boldsymbol{\sigma}} \delta^\dagger = \det \mathbf{r} \cdot \hat{\boldsymbol{\sigma}} = -\mathbf{r}^2, \tag{1.97}$$

we see that the transformation leaves $\mathbf{r}^2$ invariant and is therefore a rotation.

The composition law, which we can easily derive from the previous equations, then proves that the obtained 3-dimensional rotations from the rotation group—SO(3):

$$R(\delta_1)R(\delta_2) = R(\delta_1\delta_2). \tag{1.98}$$

However, from Eq. (1.95) it follows that we have a two-to-one homomorphism between SU(2) and SO(3):

$$\pm\delta \to R(\delta), \tag{1.99}$$

which means that both δ and $-\delta$ in the 2-dimensional spin space induce the same rotation in the 3-dimensional laboratory space.

Let us elaborate on this result. An arbitrary rotation in a 3-dimensional space can be described by three angles about coordinates axes. These are called the *Euler angles.* There are different conventions about the axes we choose. We adopt the *zyz* convention: rotation through angle ψ about the z-axis, through θ about the new y-axis, and through ϕ about the new z-axis.

If we choose $\delta_{11} = e^{i\phi/2}$ and $\delta_{12} = 0$ in Eq. (1.94), then Eq. (1.93) yields

$$x' = x\cos\phi - y\sin\phi, \quad y' = x\cos\phi + y\sin\phi, \quad z' = z, \tag{1.100}$$

which amounts to

$$\begin{bmatrix} e^{\frac{i\phi}{2}} & 0 \\ 0 & e^{-\frac{i\phi}{2}} \end{bmatrix} \longrightarrow \begin{bmatrix} \cos\phi & -\sin\phi & 0 \\ \sin\phi & \cos\phi & 0 \\ 0 & 0 & 1 \end{bmatrix} = R(\phi, 0, 0). \tag{1.101}$$

With reference to Eq. (1.93) we could have also written

$$\begin{bmatrix} x' \\ y' \\ z' \end{bmatrix} = \begin{bmatrix} \cos\phi & -\sin\phi & 0 \\ \sin\phi & \cos\phi & 0 \\ 0 & & 1 \end{bmatrix} \begin{bmatrix} x \\ y \\ z \end{bmatrix} = R(\phi, 0, 0) \begin{bmatrix} x \\ y \\ z \end{bmatrix}. \tag{1.102}$$

If we choose $\delta_{11} = \cos\theta/2$ and $\delta_{12} = \sin\theta/2$, we get

$$\begin{bmatrix} \cos\frac{\theta}{2} & \sin\frac{\theta}{2} \\ -\sin\frac{\theta}{2} & \cos\frac{\theta}{2} \end{bmatrix} \longrightarrow \begin{bmatrix} \cos\theta & 0 & \sin\theta \\ 0 & 1 & 0 \\ -\sin\theta & 0 & \cos\theta \end{bmatrix} = R(0,\theta,0). \quad (1.103)$$

Hence, a general morphism for the Euler angles $\{\phi, \theta, \psi\}$ reads

$$\begin{bmatrix} e^{\frac{i\psi}{2}} & 0 \\ 0 & e^{-\frac{i\psi}{2}} \end{bmatrix} \begin{bmatrix} \cos\frac{\theta}{2} & \sin\frac{\theta}{2} \\ -\sin\frac{\theta}{2} & \cos\frac{\theta}{2} \end{bmatrix} \begin{bmatrix} e^{\frac{i\phi}{2}} & 0 \\ 0 & e^{-\frac{i\phi}{2}} \end{bmatrix} \rightarrow R(\phi,\theta,\psi), \quad (1.104)$$

where we took the composition law

$$R(\phi,\theta,\psi) = R(\psi,0,0)R(0,\theta,0)R(\phi,0,0) \quad (1.105)$$

into account.

To sum up, Eqs. (1.83), (1.84), and (1.92) are the most general operators and matrices describing the action of an arbitrary gate on a quantum two-level system. As we have seen in Eqs. (1.100)–(1.105), these matrices can induce a corresponding description in the 3-dimensional space, but the resulting morphism is a homomorphism, not an isomorphism, as shown by Eq. (1.99). This outcome is not a problem as long as we deal with isolated qubits. Then the homomorphism reduces to two different global phases that do not make a difference for a measurement. For instance, 2π and 4π rotations in the 2-dimensional spin space give (Eq. (1.103)): $+\mathbb{1}$ and $-\mathbb{1}$, respectively, which both correspond to $+\mathbb{1}$ in the corresponding 3-dimensional space ($R(0,2\pi,0) = R(0,4\pi,0) = \mathbb{1}$). However, for a qubit entangled with another qubit, different global phases of each qubit do make a difference.

Therefore, we take SU(2) and not SO(3) as the basic group. Also, by multiplying SU(2) by itself, we can arrive at both half-integer and integer generator representations, which cannot be achieved within SO(3).

A *generator* for $R(\phi,0,0)$ is the constant matrix I_3 in the following representation of R:

$$R(\phi,0,0) = e^{-i\phi I_3}, \qquad \text{where } I_3 = \begin{bmatrix} 0 & -i & 0 \\ i & 0 & 0 \\ 0 & 0 & 0 \end{bmatrix}. \quad (1.106)$$

The matrix exponential here is defined as follows [Huang, 1999]:

$$\exp(A) = e^A = \sum_{n=1}^{\infty} \frac{A^n}{n!} = \mathbb{1} + A + \frac{AA}{2!} + \frac{AA}{2!} \cdots . \quad (1.107)$$

The series converges for any square matrix. Its implementation in Wolfram's *Mathematica* (MatrixExp) comes in handy:

$$\text{MatrixExp}[e^{-i\phi I_3}] = \begin{bmatrix} \cos\phi & -\sin\phi & 0 \\ \sin\phi & \cos\phi & 0 \\ 0 & 0 & 1 \end{bmatrix}, \tag{1.108}$$

$$e^{-i\theta I_2} = \text{MatrixExp}[-ie^{\theta}\begin{bmatrix} 0 & 0 & i \\ 0 & 0 & 0 \\ -i & 0 & 0 \end{bmatrix}] = \begin{bmatrix} \cos\theta & 0 & \sin\theta \\ 0 & 1 & 0 \\ -\sin\theta & 0 & \cos\theta \end{bmatrix}.$$

With the help of the generators I_j, we can rewrite Eq. (1.105) as

$$D^1(\phi,\theta,\psi) = e^{-i\psi I_3}e^{-i\theta I_2}e^{-i\phi I_3}. \tag{1.109}$$

Analogously, in SU(2), the left-hand side of the inducement (1.104) can be written as

$$D^{\frac{1}{2}}(\phi,\theta,\psi) = e^{-i\frac{\psi}{2}\sigma_3}e^{-i\frac{\theta}{2}\sigma_2}e^{-i\frac{\phi}{2}\sigma_3}. \tag{1.110}$$

Note that the rotations by angle θ in SO(3) and by $\theta/2$ in SU(2) correspond to the phase shift $\frac{\phi_0-\phi_1}{2}$ in Eq. (1.58) and to asymmetry of the beam splitter $t \neq r$ in Eqs. (1.69) and (1.70).

Taken together, we can represent any one-qubit operation and any quantum gate by Eqs. (1.84), (1.89), (1.92), (1.104), (1.109), and (1.110), which are all equivalent up to a global phase. A correspondence of qubit states and Euler angles in SO(3) is often presented on the so-called Bloch sphere shown in Fig. 1.20.

The following special case of the preceding correspondence is often used in the literature. Consider

$$\mathbf{r} = \{\sin\theta\cos\phi, \sin\theta\sin\phi, \cos\theta\} \tag{1.111}$$

as a choice for $\mathbf{r}$ in Eq. (1.92). Then the density matrix from the latter equation reads (we drop the "hat" over ρ):

$$\begin{aligned}
\rho &= \frac{1}{2}\begin{bmatrix} 1+\cos\theta & e^{-i\phi}\sin\theta \\ e^{+i\phi}\sin\theta & 1-\cos\theta \end{bmatrix} \\
&= \begin{bmatrix} \cos^2\theta & e^{-i\phi}\sin\theta\cos\theta \\ e^{+i\phi}\sin\theta\cos\theta & \sin^2\theta \end{bmatrix} \\
&= \cos^2\frac{\theta}{2}|0\rangle\langle 0| + e^{-i\phi}\sin\frac{\theta}{2}\cos\frac{\theta}{2}|0\rangle\langle 1| \\
&\quad + e^{i\phi}\sin\frac{\theta}{2}\cos\frac{\theta}{2}|1\rangle\langle 0| + \sin^2\frac{\theta}{2}|1\rangle\langle 1| \;=\; |\Psi\rangle\langle\Psi|,
\end{aligned} \tag{1.112}$$

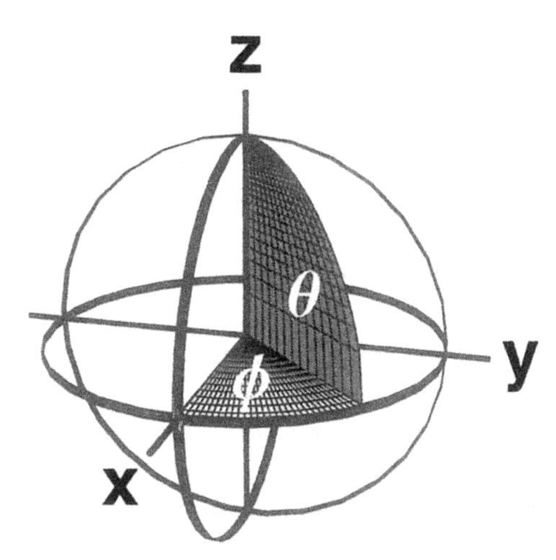

Figure 1.20. Bloch sphere. All pure states are on its surface. The $\phi = \pi/4$, $\theta = \pi/2$ case is shown. $|0\rangle$ is at $z = 1$, $|1\rangle$ at $z = -1$, $|0\rangle + |1\rangle$ at $x = 1$, $|0\rangle + i|1\rangle$ at $y = 1$, $|0\rangle - |1\rangle$ at $x = -1$, and $|0\rangle - i|1\rangle$ at $y = -1$. Mixed states are inside the sphere. For instance, $\frac{1}{2}(|0\rangle\langle 0| + |1\rangle\langle 1|)$ is at $x = y = z = 0$.

where

$$|\Psi\rangle = e^{i\chi}(\cos\frac{\theta}{2}|0\rangle + e^{i\phi}\sin\frac{\theta}{2}|1\rangle). \tag{1.113}$$

The density matrix (1.112) satisfies the condition $\rho^2 = \rho$ from Eq. (1.49) and therefore is a pure state. For the last line of Eq. (1.112), cf. Eq. (1.33). The overall phase term $e^{i\chi}$ does not go through to either ρ or to the Bloch sphere representation. Note that both θ and $\theta + 2\pi$, i.e., both $|\Psi\rangle$ and $-|\Psi\rangle$, correspond to a single Bloch sphere representation for the same reason (see Eq. (1.99)). Note also that we have already used a special case ($\phi = \chi = 0$) of the state given by Eq. (1.113) in Sec. 1.15, Fig. 1.12.

1.19 Other Qubits and Universal Gates

General N-dimensional operators $\hat{U}$ (and corresponding $N \times N$ matrices), being quantum mechanical operators, have to be unitary (Eqs. (1.38), (1.37), and (1.35)). This condition assures time reversibility of qubit operations and gates. Dimension N is determined by the number of qubits, n, since $N = 2^n$.

The ideas behind quantum computing are mostly based on two-level systems or qubits as described in Sec. 1.18—although there are other elaborations, for instance, three level systems (*qutrits*) on SU(3) [Byrd, 1998]. Usually, one such system is addressed at a time. We call it a *target* qubit. The way it will be processed is conditioned by the states the other qubits are in at the time. They are called *control* qubits. At another moment in time, these roles may interchange. Also, later on, we will see that at a particular stage of calculation the qubits need not be of either kind—they can be just a passive time line.

These two features—reversibility and control-target gate design—have obvious similarities to the reversible gates we elaborated on in Sec. 1.7. In order to use the terminology and design of n-bit reversible gates and apply them to n-qubit quantum gates, let us first establish a way to express reversible control-target design by means of unitary matrices. Then we shall explain why this formalism, borrowed from the terminology of reversible gates, works with quantum gates.

The formalism of n-qubit gates exploits and extends two main features of reversible gates (presented in Sec. 1.7): (a) reversibility—implemented as reversibility in time through the requirement that operators and matrices be unitary—and (b) the control-target model of building circuits.

A target is a single qubit, and a gate either acts on it as described by a general 2×2 unitary matrix, provided that all the control qubits are in the state $|1\rangle$, or leaves it unchanged if they are not:

$$|\psi_n{}'\rangle = \begin{cases} \hat{U}|\psi_n\rangle & \text{for } |\psi_1\rangle = |1\rangle, \ldots, |\psi_{n-1}\rangle = |1\rangle \\ |\psi_n\rangle & \text{otherwise,} \end{cases} \tag{1.114}$$

where $|\psi_j\rangle$, $j = 1, \ldots, n$, can be equal to $|0\rangle$, to $|1\rangle$, or to $\alpha_j|0\rangle + \beta_j|1\rangle$. To match such a description of a gate, its matrix must consist of $2^n/2-1$ 2×2 unit matrices as shown in Fig. 1.21 (p. 52).

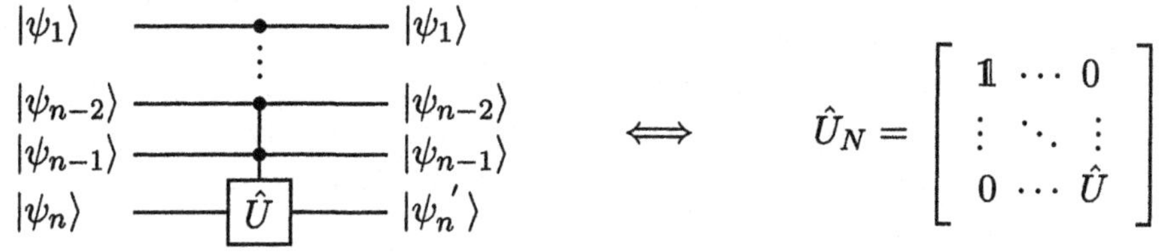

$$\Longleftrightarrow \qquad \hat{U}_N = \begin{bmatrix} \mathbb{1} & \cdots & 0 \\ \vdots & \ddots & \vdots \\ 0 & \cdots & \hat{U} \end{bmatrix}$$

Figure 1.21. An n-qubit circuit diagram is equivalent to an $N \times N$ matrix $\hat{U}_N$, where $N = 2^n$. $|\psi_1\rangle, \ldots, |\psi_{n-1}\rangle$ are control qubits and $|\psi_n\rangle$ is the target qubit. The diagonal has $n-1$ 2×2 unit matrices, $\mathbb{1}$, and a general 2-dimensional 2×2 unitary matrix, $\hat{U}$.

It is instructive to compare Eq. (1.114) and Fig. 1.21 with Eq. (1.7) and give the conditional in Eq. (1.114) yet another explanation. If we write down n-qubit vectors as

$$|\underbrace{0\ldots0}_{2^n \text{ times}}\rangle = \begin{bmatrix} 1 \\ 0 \\ \vdots \\ 0 \end{bmatrix}, \quad |\underbrace{0\ldots0}_{2^n-1 \text{ times}}1\rangle = \begin{bmatrix} 0 \\ 1 \\ \vdots \\ 0 \end{bmatrix}, \ldots, |\underbrace{1\ldots1}_{2^n \text{ times}}\rangle = \begin{bmatrix} 0 \\ 0 \\ \vdots \\ 1 \end{bmatrix}, \tag{1.115}$$

we can see that $\hat{U}_N$ will never change any of the first 2^n-2 vectors, and we obtain $\mathbb{1}$s. $\hat{U}_N$ can only change—through 2×2 $\hat{U}$—the last two vectors.

Let us now explain why this formalism works with quantum gates. The idea is to express a general matrix $\hat{G}$ by means of gates analogous to $\hat{U}_N$ presented in Fig. 1.21. The following theorem provides us with the explanation.

THEOREM 1.3 "Two-level gates are universal." *Any $2^n \times 2^n$ unitary matrix $\hat{U}^{(2^n)}$ can be expressed as a product of matrices that act nontrivially only on two vector components.*

Proof Consider the following $N-1$ ($N=2^n$) products, where the second matrix is always the same unitary matrix $\hat{U}^{(N)}$ and the first one is $\hat{\alpha}_j^{(N)}$ with $j=2,\ldots,N$:

$$\begin{bmatrix} \alpha_{11} & 0 & \cdots & \alpha_{1j} & \cdots & 0 \\ 0 & \alpha_{22} & \cdots & 0 & \cdots & 0 \\ \vdots & \vdots & \ddots & \vdots & & \vdots \\ \alpha_{j1} & 0 & \cdots & \alpha_{jj} & \cdots & 0 \\ \vdots & \vdots & & \vdots & \ddots & \vdots \\ 0 & 0 & \cdots & 0 & \cdots & \alpha_{NN} \end{bmatrix} \begin{bmatrix} u_{11} & u_{12} & \cdots & u_{1j} & \cdots & u_{1N} \\ u_{21} & u_{22} & \cdots & u_{2j} & \cdots & u_{2N} \\ \vdots & \vdots & \ddots & \vdots & & \vdots \\ u_{j1} & u_{j2} & \cdots & u_{jj} & \cdots & u_{NN} \\ \vdots & \vdots & & \vdots & \ddots & \vdots \\ u_{N1} & u_{N2} & \cdots & u_{Nj} & \cdots & u_{NN} \end{bmatrix}, \quad (1.116)$$

where

$$\begin{aligned} \alpha_{11} &= \frac{u_{jj}}{\sqrt{|u_{1j}|^2+|u_{jj}|^2}}, \qquad \alpha_{1j} = \frac{-u_{1j}}{\sqrt{|u_{1j}|^2+|u_{jj}|^2}}, \\ \alpha_{j1} &= \alpha_{1j}^*, \qquad \alpha_{jj} = -\alpha_{11}^* \\ a_{22} = \cdots = a_{j-1j-1} = a_{j+1j+1} \cdots = a_{NN} &= 1 \qquad \text{for} \quad j \neq 2, N \\ a_{33} = \cdots = a_{NN} &= 1 \qquad \text{for} \quad j = 2 \\ a_{22} = \cdots = a_{N-1N-1} &= 1 \qquad \text{for} \quad j = N, \end{aligned} \quad (1.117)$$

The element in the jth column of the first row of the product matrix from Eq. (1.116) is therefore equal to 0. Also, since $\hat{U}^{(N)}$ is unitary and therefore $u_{ji}=u_{ij}^*$, the element in the jth row of the first column of this product matrix is equal to 0. The first matrices in Eq. (1.116), i.e., $\hat{\alpha}_j^{(N)}$, are, given the condition (1.117), unitary: $\hat{\alpha}_j^{(N)}\alpha_j^{(N)\dagger}=\mathbb{1}$, where $\mathbb{1}$ is an $N \times N$ unit matrix. Since $U^{(N)}$ is also a unitary matrix, so is $\alpha_j^{(N)}U^{(N)}$.

Now, $U^{(N-1)} = \hat{\alpha}_{2^n}^{(N)}\hat{\alpha}_{2^n-1}^{(N)}\cdots\hat{\alpha}_2^{(N)}U^{(N)}$ is a matrix that has all the elements in the first row equal to 0, except the first one from the first column (which is 1). This matrix must also be unitary, since $\alpha_j^{(N)}U^{(N)}$ is unitary for any j, $j = 2, 3, \ldots, 2^n$. So we get

$$U^{(N-1)} = \hat{\alpha}_{2^n}^{(N)}\hat{\alpha}_{2^n-1}^{(N)}\ldots\hat{\alpha}_2^{(N)}U^{(N)} = \begin{bmatrix} 1 & 0 & \cdots & 0 & \cdots & 0 \\ 0 & u'_{22} & \cdots & u'_{2j} & \cdots & u'_{2N} \\ \vdots & \vdots & \ddots & \vdots & & \vdots \\ 0 & u'_{j2} & \cdots & u'_{jj} & \cdots & u'_{NN} \\ \vdots & \vdots & & \vdots & \ddots & \vdots \\ 0 & u'_{N2} & \cdots & u'_{Nj} & \cdots & u'_{NN} \end{bmatrix}. \quad (1.118)$$

We repeat our procedure on $U^{(N-2)}$, $U^{(N-3)}, \ldots$ to obtain

$$\begin{aligned} U^{(N-2)} &= \hat{\alpha}_{2^n}^{(N-1)}\ldots\hat{\alpha}_3^{(N)}U^{(N-1)} = \alpha_{2^n}^{(N-2)}\ldots\hat{\alpha}_2^{(N)}U^{(N)} \\ U^{(N-3)} &= \hat{\alpha}_{2^n}^{(N-3)}\ldots\hat{\alpha}_2^{(N)}U^{(N)} \\ \vdots &= \cdots \\ \mathbb{1} = U^{(1)} &= \hat{\alpha}_{2^n}^{(1)}\ldots\hat{\alpha}_2^{(N)}U^{(N)}, \end{aligned} \quad (1.119)$$

from which we get

$$U^{(N)} = \hat{\alpha}_{2^n}^{(1)\dagger}\ldots\hat{\alpha}_2^{(N)\dagger}. \quad (1.120)$$

There are $(N-1) + \cdots + 1 = N(N-1)/2 = 2^n(2^n-1)/2$ such α's. They all act nontrivially only on at most two rows of any N-row single column matrices, which proves the claim of the theorem. ∎

This result enables us to reduce any unitary gate to a chain of cascaded controlled gates, denoted controlled-controlled-. . . -U or controlled^n-U, and therefore we should engage the gates for each particular qubit within a specific time window. Applications, however, require that the number of qubit gates that are engaged within a time window be reduced to a minimum. Again, we can learn from the reversible gate approach. In Sec. 1.7, we said that there are many 3-bit universal reversible gates. We can show a similar result for quantum gates. For example, we can substitute controlled^4-U with a sequence of Toffoli gates, as shown in Fig. 1.22.

Here we reach the limits of comparing quantum with reversible circuits and come to a point where quantum circuits essentially surpass reversible ones. Three-bit universal gates—for example, the Toffoli gate—are the smallest universal reversible gates [Toffoli, 1980], while—as follows from Theorem 1.3 and the elaboration below—almost any two-qubit quantum gate is universal.

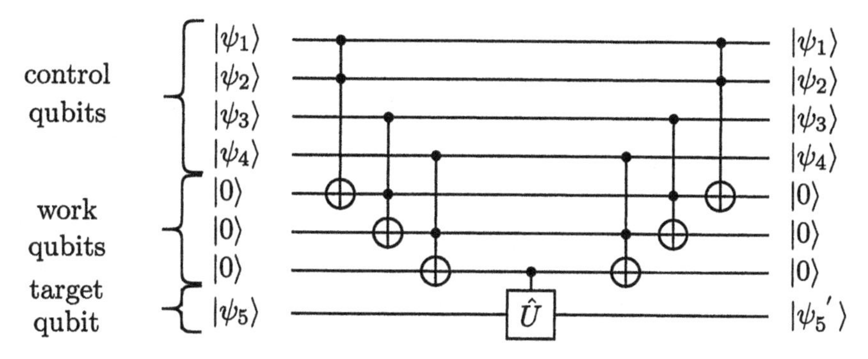

Figure 1.22. A 5-qubit controlled^4-U gate of the type shown in Fig. 1.21 is here implemented by means of three Toffoli gates. When, for instance, $|\psi_i\rangle = |1\rangle$, $i = 1,\ldots,4$, the first and second Toffoli gates change the states of the first and second work qubit from $|0\rangle$ to $|1\rangle$. The third Toffoli gate changes the third work qubit into control qubit $|1\rangle$ for the target qubit gate U. In general, control qubits are in a superposition state, as for example in Figs. 1.24 (p. 57) and 1.28 (p. 77).

To see this and at the same time to better understand the correspondence between quantum circuit diagrams and unitary operators, let us consider the example presented in Fig. 1.23. In this figure, we take a controlled-controlled-$\hat{U}$ gate and express it by means of several CNOT and controlled-$\hat{V}$ gates, where $\hat{V}^2 = \hat{U}$ ($\hat{V}$ is, of course, also unitary). Note that the Toffoli gate is a special case of the controlled-controlled-$\hat{U}$ gate.

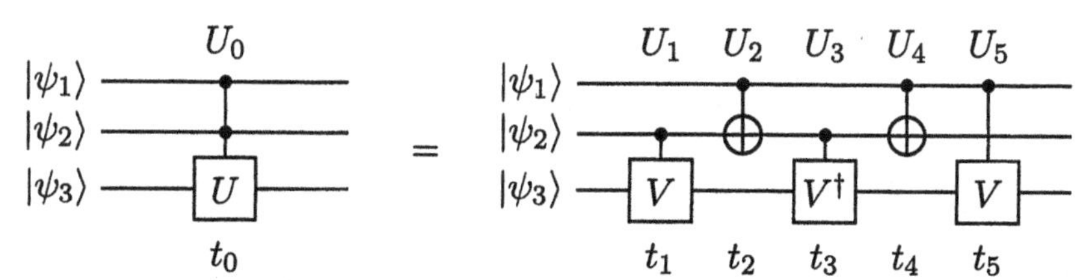

Figure 1.23. The 3-qubit gate controlled^2-U expressed by means of five 2-qubit gates. U_j, $j = 1,\ldots,5$, are the matrices describing the gates. t_j, $j = 1,\ldots,5$, are the corresponding times. $V^2 = U$.

Every quantum gate presented in a quantum circuit diagram corresponds to a unitary matrix defining it. So in Fig. 1.23, matrix U_0 is given as

$$\begin{bmatrix} \mathbb{1} & & \cdots & 0 \\ & \mathbb{1} & & \vdots \\ \vdots & & \mathbb{1} & \\ 0 & \cdots & & U \end{bmatrix}, \tag{1.121}$$

where the $\mathbb{1}$'s are 2×2 unit vectors and U is a 2×2 single-qubit unitary gate matrix.

To determine U_j, $j = 1, \ldots, 5$, we shall first reconsider the correspondence between qubit vectors and their column matrices from Eq. (1.115). The basis vectors of the first qubit, are $|0000\rangle$ and $|0001\rangle$ and in the one-column matrix representation, these are the ones with 1s in the first and second row as in the first two matrices of Eq. (1.115). The state $|1\rangle$ of the second qubit we find in the third, fourth, seventh, and eighth ket ($|010\rangle$, $|011\rangle$, $|110\rangle$, and $|111\rangle$, respectively). This means that the one-column matrices have 1s in the third, fourth, seventh, and eighth row of the the qubit matrices, and that we have to have one 2×2 matrix acting on the third and fourth row of the latter one-column matrices and another on the seventh, and eighth row. In a similar way, we determine the other U_j, $j = 2, \ldots, 5$, matrices so as eventually to have

$$U_1 = \begin{bmatrix} \mathbb{1} & & & 0 \\ & V & & \\ & & \mathbb{1} & \\ 0 & & & V \end{bmatrix}, U_2 = \begin{bmatrix} \mathbb{1} & & & 0 \\ & \mathbb{1} & & \\ & & 0 & \mathbb{1} \\ 0 & & \mathbb{1} & 0 \end{bmatrix}, U_3 = \begin{bmatrix} \mathbb{1} & & & 0 \\ & V^\dagger & & \\ & & \mathbb{1} & \\ 0 & & & V^\dagger \end{bmatrix},$$
$$U_4 = \begin{bmatrix} \mathbb{1} & & & 0 \\ & \mathbb{1} & & \\ & & 0 & \mathbb{1} \\ 0 & & \mathbb{1} & 0 \end{bmatrix}, \quad \text{and} \quad U_5 = \begin{bmatrix} \mathbb{1} & & & 0 \\ & \mathbb{1} & & \\ & & V & \\ 0 & & & V \end{bmatrix}. \tag{1.122}$$

Now we simply have to check that

$$U_5 U_4 U_3 U_2 U_1 = U_0 \tag{1.123}$$

to prove that any quantum gate can be expressed by means of two-qubit controlled gates.

Since, as we have shown in Sec. 1.18 (p. 45), single-qubit gates (unitary operators) can be obtained from each other by rotations, any quantum gate can consequently be expressed by means of CNOTs and single gates. CNOTs allow us to submit target qubits to control qubits, and single qubit gates allow us to rotate target states to desired ones.

1.20 Teleportation of Copies and the No-Cloning Theorem

The presentation of entangled states using a standard circuit diagram is not straightforward because the formalism of the diagrams is borrowed from reversible gate notation, and the controlled gates there assume definite bits on each input and output. In a two- or more qubit quantum circuit, control and/or target qubits can be in an unknown state of superposition of basis vectors. The outputs we get must not be treated independently at the exits of the gates, because qubits appear entangled, and only in an entangled state can they give us the time benefit we expect of a quantum computer. Hence we often treat the exits of controlled gates as shown in Fig. 1.23, although many times it turns out that writing the outgoing states onto each line as shown in Figs. 1.28 (p. 77) and 1.30 (p. 79) (pp. 77, 79), is unavoidable if one wants to keep the circuit diagrams tidy. In the latter case, we have to keep in mind that the lines do not correspond to single qubits but rather to all the entangled qubits together. Hence when we use one of them, the others instantaneously change as in Fig 1.28 (p. 77)).

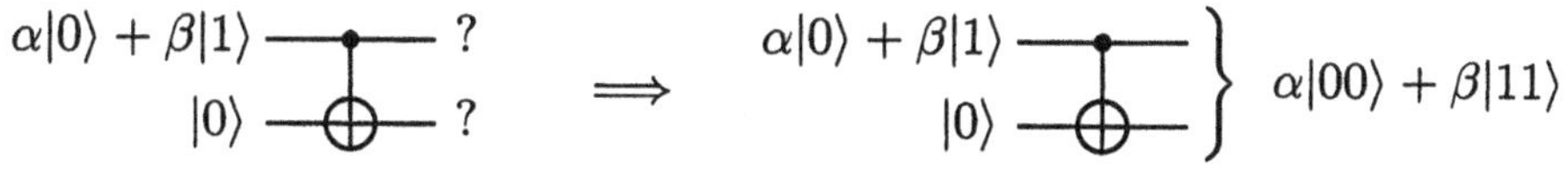

$\alpha|0\rangle + \beta|1\rangle$ — ? $\quad\Longrightarrow\quad$ $\alpha|0\rangle + \beta|1\rangle$ —
$|0\rangle$ — ? $\qquad\qquad\qquad\;$ $|0\rangle$ — $\Big\}\ \alpha|00\rangle + \beta|11\rangle$

Figure 1.24. When the control qubit of CNOT is in a superposed state, it does not fit into the standard circuit diagram smoothly when it entangles with another qubit. One of the solutions is a curly bracket, which signifies that the outgoing qubits are entangled and that neither of them is in a definite state of its own. The other is shown in Figs. 1.28 and 1.30 (pp. 77, 79) in Sec. 1.22 (p. 72).

A formal way of writing down the circuit in Fig. 1.24 is

$$\begin{aligned}&\mathrm{CNOT}(\alpha|0\rangle + \beta|1\rangle)|0\rangle = \\ &\alpha\,\mathrm{CNOT}\,|0\rangle|0\rangle + \beta\,\mathrm{CNOT}\,|1\rangle|0\rangle = \alpha\,|00\rangle + \beta\,|11\rangle. \end{aligned} \tag{1.124}$$

The quantum circuit formalism considers superposed states $|0\rangle$ and $|1\rangle$ in the control qubit to be "Schrödinger cat states." For $|0\rangle$ leaves the target $|0\rangle$ unchanged to form the 2-qubit state $|00\rangle$, and $|1\rangle$ changes it into $|1\rangle$ to form $|11\rangle$. Both obtained states are, however, "virtual" and make sense only when we treat them together as entangled state $\alpha|00\rangle + \beta|11\rangle$.

In Sec. 1.19 (p. 51), we have shown that any quantum gate can be expressed by means of CNOTs and single gates and that they allow us to rotate target states to arbitrary new states. Therefore, by using CNOTs, we can arbitrarily rotate a qubit into any desired new position. However, we cannot start with a nontrivial superposition, also called the *unknown state,*[12]

$$\frac{1}{\sqrt{|\alpha|^2+|\beta|^2}}(\alpha|0\rangle+\beta|1\rangle), \qquad \alpha,\beta\neq 0,1 \tag{1.125}$$

and a "known state," either $|\psi\rangle=|0\rangle$ or $|\psi\rangle=|1\rangle$, so as to arrive at two replicas of the unknown state:

$$(\alpha|0\rangle+\beta|1\rangle)\otimes|\psi\rangle \;\not\Rightarrow\; (\alpha|0\rangle+\beta|1\rangle)\otimes(\alpha|0\rangle+\beta|1\rangle). \tag{1.126}$$

In general the following *no-cloning theorem* holds.

THEOREM 1.4 *Unknown quantum states cannot be cloned.*

Proof Let us assume that there exists a cloning operator $\hat{C}$. By linearity we have

$$\hat{C}(\alpha|0\rangle+\beta|1\rangle)=\alpha\hat{C}|0\rangle+\beta\hat{C}|1\rangle. \tag{1.127}$$

The left-hand side of Eq. (1.127):

$$(\alpha|0\rangle+\beta|1\rangle)\otimes(\alpha|0\rangle+\beta|1\rangle)=\alpha^2|00\rangle+\alpha\beta(|10\rangle+|01\rangle)+\beta^2|11\rangle$$

is, however, not equal to its right-hand side

$$\alpha\hat{C}|0\rangle+\beta\hat{C}|1\rangle=\alpha|00\rangle+\beta|11\rangle.$$

■

To understand the difference between a qubit and a bit with respect to this qubit feature, let us compare Fig. 1.23 (p. 55) with the classical reversible CNOT gate shown in Fig. 1.4 (p. 16). The latter CNOT for arbitrary control bit **A** and the target bit **0** gives two **A**s. We might be tempted to consider this arbitrary control bit "unknown." However, it is always "known" because it is either 0 or 1. This is why classical processors—reversible as well standard ones—can be based on bit replicas.

We cannot clone unknown qubits, but we can nevertheless copy them. We just have to pay a price for doing so: the original must be destroyed in the process. To see why, we have to analyze the entanglement from yet another angle, which is essential for understanding its role in teleportation and computation. Let us take a look at the experimental proposal shown in Fig. 1.25 [Pavičić and Summhammer, 1994].

[12] We say "unknown" because when we measure it we randomly get either a $|0\rangle$-click or a $|1\rangle$-click, although in the long (N) run we get $N|\alpha|^2/(|\alpha|^2+|\beta|^2)$ $|0\rangle$-clicks and $N|\beta|^2/(|\alpha|^2+|\beta|^2)$ $|1\rangle$-clicks.

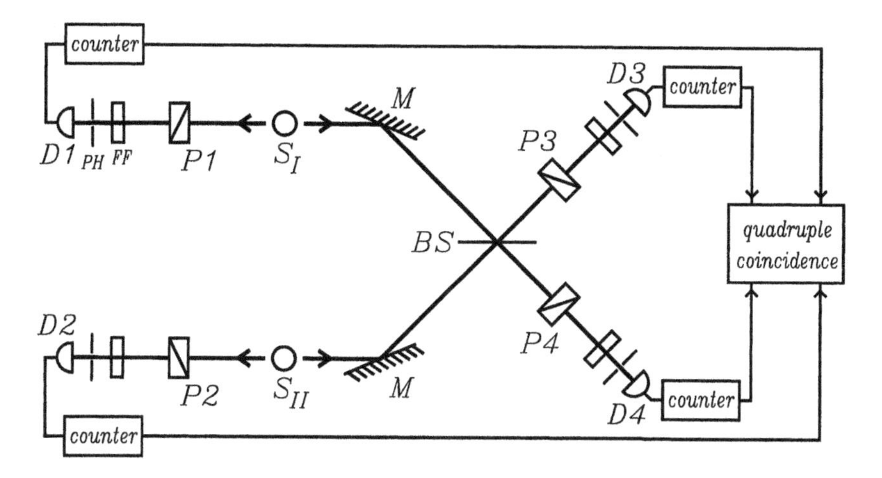

Figure 1.25. Figure taken from [Pavičić and Summhammer, 1994]: "Two photons from different unpolarized sources each pass through a polarizer to a detector. Although their trajectories never mix or cross they exhibit 4th-order-interference-like correlations when the other two photons interfere at a beam splitter even when the latter two do not pass any polarizers at all." [Pavičić and Summhammer, 1994] PH are pinholes, FF frequency filters, P polarizers, D detectors, and BS is a beam splitter.

Two independent sources, S_I and S_{II}, both simultaneously emit two photons correlated in polarization to the left and right. We measure the polarizations of the photons on the left with the polarization filters P1 and P2 and the photons on the right with P3 and P4. To point out that we get photons genuinely unprepared, we emphasize that the sources can in principle be atoms exhibiting cascade emission. But if we wanted a feasible experiment, we would most likely use down-converted photons [Pavičić, 1995].

The state of the four photons immediately after leaving their triplet-pair sources is described by the product of two entangled states (cf. Eq. (1.68))

$$|\Psi\rangle = \frac{1}{\sqrt{2}}(|0\rangle_1|0\rangle_3 + |1\rangle_1|1\rangle_3) \otimes \frac{1}{\sqrt{2}}(|0\rangle_2|0\rangle_4 + |1\rangle_2|1\rangle_4). \quad (1.128)$$

Here, $|0\rangle$ and $|1\rangle$ denote mutually orthogonal photon states. For example, $|0\rangle_1$ means the state of photon 1 leaving source S_I to the left, polarized in direction x. As in Sec. 1.17 we use the annihilation operator formalism. (The reader is not expected to know the formalism but only its "rules of thumb" as presented in Sec. 1.17—in the paragraphs following Eq. (1.68).) When P1 is oriented at some angle θ_1, its action (filtering) and detection by detector D1 are represented by the operator $\hat{a}_1 = \hat{a}_{01}\cos\theta_1 + \hat{a}_{11}\sin\theta_1$. The phase that the photon accumulates between the source S_I and the detector D1 is $e^{i\omega_1(r_1/c+t_0^I-t_1)}$, where ω_1 is the frequency of photon 1, r_1 is the path length from S_I to D1, c is the speed of light, t_0^I is the time of emission of a pair of photons at S_I, and t_1 is the time of detection at D1.

To describe the detection of a photon by detector D1 we apply the operator

$$\hat{E}_1 = (\hat{a}_{01}\cos\theta_1 + \hat{a}_{11}\sin\theta 1)e^{i\omega_1(r_1/c+t_0^I-t_1)} \tag{1.129}$$

to the initial state of Eq. (1.128). Similarly, the detection of photon 2 at D2 means we apply

$$\hat{E}_2 = (\hat{a}_{02}\cos\theta_2 + \hat{a}_{12}\sin\theta_2)e^{i\omega_2(r_2/c+t_0^{II}-t_2)}, \tag{1.130}$$

where the symbols are defined analogously.

On the right side of the sources, a detection at D3 can be caused by the emission of photon 3 by source S_I or the emission of photon 4 by source S_{II}. The beam splitter BS may have polarization transmittances and reflectances, denoted by T_0, T_1, and R_0, R_1, respectively. The angle of the polarizer P3 is given by θ_3. The operator describing a photon arriving at the detector D3 is

$$\begin{aligned}\hat{E}_3 &= \left(\hat{a}_{04}\sqrt{T_0}\cos\theta_3 + \hat{a}_{14}\sqrt{T_1}\sin\theta_3\right)e^{i\alpha} \\ &\quad + i\left(\hat{a}_{03}\sqrt{R_0}\cos\theta_3 + \hat{a}_{13}\sqrt{R_1}\sin\theta_3\right)e^{i\beta} \\ &= (\hat{a}_{04}A_{04} + \hat{a}_{14}A_{14})e^{i\alpha} + i(\hat{a}_{03}A_{03} + \hat{a}_{13}A_{13})e^{i\beta},\end{aligned} \tag{1.131}$$

where

$$\alpha = \omega_4(\frac{r_{II}+r_3}{c} + t_0^{II} - t_3), \qquad \beta = \omega_3(\frac{r_I+r_3}{c} + t_0^I - t_3), \tag{1.132}$$

where r_I and r_{II} denote the distances from the respective sources to BS, and r_3 and t_3 denote the distance from BS to D3 and the time of detection at D3, respectively. The meaning of the A's is obvious. For D4, one defines $\hat{E}_4$ analogously [Pavičić and Summhammer, 1994]:

$$\begin{aligned}\hat{E}_4 &= \left(\hat{a}_{03}\sqrt{T_0}\cos\theta_4 + \hat{a}_{13}\sqrt{T_1}\sin\theta_4\right)e^{i\gamma} \\ &\quad + i\left(\hat{a}_{04}\sqrt{R_0}\cos\theta_4 + \hat{a}_{14}\sqrt{R_1}\sin\theta_4\right)e^{i\delta} \\ &= (\hat{a}_{03}B_{03} + \hat{a}_{13}B_{13})e^{i\gamma} + i(\hat{a}_{04}B_{04} + \hat{a}_{14}B_{14})e^{i\delta},\end{aligned} \tag{1.133}$$

where

$$\gamma = \omega_3(\frac{r_I+r_4}{c} + t_0^I - t_4), \qquad \delta = \omega_4(\frac{r_{II}+r_4}{c} + t_0^{II} - t_4). \tag{1.134}$$

The meaning of the B's is obvious.

Up to this point in the calculation, everything has followed standard quantum mechanics: no entanglement has appeared in the sense that, as long as we keep to the "wholeness" (Niels Bohr's expression) of the experimental arrangement, a corresponding "complete" wave function of several quantum systems taking part in the experiment can always be described by a tensor product of its parts. Entanglement comes at the stage when we want to make some measurements on some subsystems and *not* some other measurements on some other subsystems, that is, when we decide to manipulate our subsystems with the aim of constructing a new quantum mechanical reality. Then—as a consequence—we also manipulate the standard formalism so as to extract the parts we need and disregard those that we do not need. These parts are determined by the requirements of quantum circuits designed to calculate and communicate. Eventually we formalize these quantum circuit requirements and rules and call them *quantum logic.* All the theoretical results supporting quantum computation and quantum communication are then called *quantum information theory.*

To arrive at this extraction from the standard formalism and use it to build quantum logic formalism, we consider the experiment presented in Fig. 1.25 (p. 59). We consider the coincidence probability for all four photons detected by detectors D1, D2, D3, and D4 (see Fig. 1.26, p. 64) only (which means not by detectors $\mathrm{D1}^{\perp}$, $\mathrm{D2}^{\perp}$, $\mathrm{D3}^{\perp}$, and $\mathrm{D4}^{\perp}$, not the photons 3, 4 together by one detector D3, etc.—we disregard all the corresponding electric-field operators). This probability reads [Pavičić and Summhammer, 1994]

$$P(\theta_1,\theta_2,\theta_3,\theta_4) = \langle\Psi|\hat{E}_1^\dagger\hat{E}_2^\dagger\hat{E}_3^\dagger\hat{E}_4^\dagger\hat{E}_4\hat{E}_3\hat{E}_2\hat{E}_1|\Psi\rangle. \quad (1.135)$$

Eqs. (1.128), (1.129), and (1.130) yield

$$\hat{E}_2\hat{E}_1|\Psi\rangle = \frac{e^{i\epsilon}}{2}\left(\cos\theta_1|0\rangle_3 + \sin\theta_1|1\rangle_3\right)\otimes\left(\cos\theta_2|0\rangle_4 + \sin\theta_2|1\rangle_4\right), \quad (1.136)$$

where the meaning of ϵ is obvious.

Applying $\hat{E}_4\hat{E}_3$ as given by Eqs. (1.131) and (1.133) to Eq. (1.136) and using rules for annihilation operators given on p. 41 we obtain

$$\hat{E}_4\hat{E}_3\hat{E}_2\hat{E}_1|\Psi\rangle = \frac{1}{2}\left[Q_{114}Q_{123}e^{i(\alpha+\gamma)} - Q_{224}Q_{213}e^{i(\beta+\delta)}\right]e^{i\epsilon}|\varnothing\rangle, \quad (1.137)$$

where $|\varnothing\rangle$ is defined on p. 41 and where

$$Q_{ijk} = \sqrt{Q_{i0}}\cos\theta_j\cos\theta_k + \sqrt{Q_{i1}}\sin\theta_j\sin\theta_k, \qquad i,j,k = 1,2, \quad (1.138)$$

where $Q_{10}, Q_{11}, Q_{10}, Q_{11}$ are T_0, T_1, R_0, R_1, respectively. Eq. (1.135) yields

$$P(\theta_1,\theta_2,\theta_3,\theta_4) = \frac{1}{4}\left[(Q_{114}Q_{123})^2 + (Q_{224}Q_{213})^2 - 2\cos(\alpha+\gamma-\beta-\delta)Q_{114}Q_{123}Q_{224}Q_{213}\right]. \quad (1.139)$$

Assuming $r_I = r_{II}$, $r_3 = r_4$, $\omega_3 = \omega_4$, and $T_0 = T_1 = R_0 = R_1 = 1/2$, we get the coincidence probability

$$P(\theta_1,\theta_2,\theta_3,\theta_4) = \tfrac{1}{16}\sin^2(\theta_1-\theta_2)\sin^2(\theta_3-\theta_4). \quad (1.140)$$

When no polarization is measured on photons 3 and 4, we get

$$P(\theta_1,\theta_2,\infty,\infty) = \frac{1}{8}\sin^2(\theta_1-\theta_2). \quad (1.141)$$

> The probability given by [Eq. (1.141)] and describing coincidence detections by D1 and D2 corresponds — when multiplied by 4 — to the following *singlet state*:
>
> $$|\Psi_s\rangle = \frac{1}{\sqrt{2}}(|0\rangle_1|1\rangle_2 - |1\rangle_1\,|0\rangle_2). \quad (1.142)$$
>
> Multiplication by 4 is for photons that emerge from the same side of BS and which we therefore dropped from our statistics [Pavičić, 1995].

This is exactly what we call the polarization *entanglement* of photons 1 and 2,

> which did not in any way directly interact and on distant distant pairs of which polarization has not been measured at all ... [and whose] trajectories never mix or cross [Pavičić and Summhammer, 1994].

The result has been verified experimentally [Pan et al., 1998].

Analogously, the probability of coincidental detection by D1 and D2$^\perp$,

$$P(\theta_1,\theta_2^\perp,\infty,\infty) = \frac{1}{8}\cos^2(\theta_1-\theta_2), \quad (1.143)$$

corresponds to the following triplet[13] state:

$$|\Psi_t\rangle = \frac{1}{\sqrt{2}}(|0\rangle_1|0\rangle_2 + |1\rangle_1\,|1\rangle_2). \quad (1.144)$$

[13]The three states $(|0\rangle_1|1\rangle_2 + |1\rangle_1\,|0\rangle_2)/\sqrt{2}$, $(|0\rangle_1|0\rangle_2 \pm |1\rangle_1\,|1\rangle_2)/\sqrt{2}$ are called the *triplet states*. They belong to the so-called *Bell states*. See Eq. (3.40), p. 154].

We see that Eq. (1.144) is actually the state from Fig. 1.24 (p. 57) with $|0\rangle_1|0\rangle_2 = |0\rangle|0\rangle$ and $|1\rangle_1|1\rangle_2 = |1\rangle|1\rangle$. Note that, to obtain Eq. (1.144) we had to multiply the corresponding substate of the overall system by 4 to get it out of the statistics of the whole system given by Eq. (1.143). And while for measurements corresponding to Eqs. (1.77) and (1.144) we do have entanglement, for other measurements in the considered setup we do not have entanglement. For example, the overall probability of detecting both photons 3, 4 in one arm of BS and detecting photons 1, 2 by D1 and D2 is given by

$$\begin{aligned} P(\theta_1,\ \theta_2,\ \theta_3\times\theta_4) &= \langle\Psi|\hat{E}_1^\dagger\hat{E}_2^\dagger\hat{E}_3^\dagger\hat{E}_3^\dagger\hat{E}_3\hat{E}_3\hat{E}_2\hat{E}_1|\Psi\rangle \\ &= \frac{1}{16}\left[\cos(\theta_1-\theta_3)\cos(\theta_2-\theta_3)+\cos(\theta_1-\theta_4)\cos(\theta_2-\theta_4)\right]^2, \end{aligned}$$

which for removed polarizers P3 and P4 reads

$$P(\theta_1,\theta_2,\infty\times\infty) = \frac{1}{8}[1+\cos^2(\theta_1-\theta_2)]\,.$$

We can also see that by removing one of the polarizers P1 and P2, say P2, we lose any left–right (Bell-like) spin correlation completely: $P(\theta_1,\infty,\theta_3,\theta_4)=\frac{1}{8}\sin^2(\theta_3-\theta_4)$, $P(\theta_1,\infty,\infty,\infty)=\frac{1}{4}$, $P(\theta_1,\infty,\infty\times\infty)=\frac{1}{4}$ [Pavičić, 1995]. Hence, the entanglement is just a property of some subsystems of the whole composite system under a particular measurement arrangement.

If we substitute the following product of singlet states

$$|\Psi\rangle = \frac{1}{\sqrt{2}}\left(|0\rangle_1|1\rangle_3\ -\ |1\rangle_1|0\rangle_3\right)\otimes\frac{1}{\sqrt{2}}\left(|0\rangle_2|1\rangle_4\ -\ |1\rangle_2|0\rangle_4\right) \quad (1.145)$$

for the triplet sources given by Eq. (1.128), we get exactly the same entanglement as above, i.e., the state (1.142) and the probability (1.141). This outcome reveals an entanglement of photons from such pairs as almost synonymous with *teleportation.*

To see this, let us look at source S1 in Fig. 1.26. (Sources S1 and S2 are simultaneously triggered by a common pumping laser beam.) Photons coming out of source S1 are in the singlet state, and therefore their polarizations are completely unprepared but correlated. With such an unprepared polarization, one of the photons from source S1 (photon 2) arrives at beam splitter BS, interferes there with another photon coming from source S2, loses its polarization and "teleports" that polarization to the second photon from source S2, i.e., to photon 4. What does this mean? It means that by measuring the polarization of photon 4, we recover the polarization of the photon coming from source S1 to beam splitter BS. How do we know this? By measuring polarization

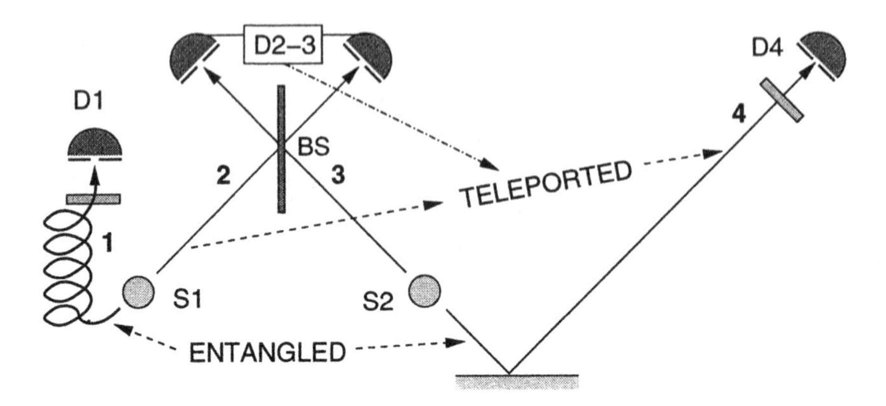

Figure 1.26. Figure according to [Pavičić, 1995]: "In the experiment, two photons from two singlets interfere at a beam splitter, and as a result the other two photons—which nowhere interacted and whose paths nowhere crossed—exhibit a 100% correlation in polarization, even when no polarization has been measured in the first two photons." [Pavičić, 1995]

of photon 1 by detector D1. (Since the photons coming out of source S1 are in the singlet state, measuring the polarization of photon 1 answers the question of which polarization the other photon coming out of source S1 should have had if it had been measured directly. The same holds for the triplet states.) This outcome has also been verified experimentally [Bouwmeester et al., 1997]. The experiment actually confirms Eq. (1.141). So both entanglement and teleportation are about engineering particular subsystems with particular properties corresponding to just some parts of a complete mathematical description of the whole system.

1.21 Quantum Cryptography

Quantum communication theory, which has been developed hand in hand with quantum computation theory, is also a part of quantum information theory and includes quantum cryptography, quantum networks, quantum buses, and quantum repeaters. In this section we shall elaborate on quantum cryptography to round out the points we put forward in Sec. 1.15 (p. 31). There, we mentioned that the development of quantum cryptography has been supported by results in quantum computation theory, in particular by Shor's algorithm.

Actually, quantum cryptography comes as a remedy to an already rising paranoia about possible failures of classical cryptography. The concern is not only that someone may find a polynomially complex classical algorithm for factoring large numbers and thereby, overnight, leave

Internet without protection, but also that in ten or twenty years' time one could decipher (using today's algorithms and the assumed computer speedup within this period) all the documents one eavesdrops and stores today. Hence, a feasible cryptography that would make eavesdropped documents unbreakable forever is a highly sought-after goal. Quantum cryptography promises to fulfill this dream. To better understand how, we shall first review the relevant points of classical cryptography.

There are three types of classical cryptographic algorithms:

- Private (secret) key (symmetric) algorithms—single key for both encryption and decryption;
- Public key (asymmetric) algorithms—one key for encryption and another for decryption;
- One-way algorithms (hash functions)—no key; irreversible encryption.

Hash functions are used to ensure the integrity of systems (to detect unauthorized changes of files and to encrypt passwords), and they are therefore not our concern here.

Private key cryptography is *the* classical cryptography known for four thousand years and provides ways of encrypting messages by a sender, often called *Alice*, with a key and decrypting the messages by a receiver, often called *Bob*, with the same key. This kind of cryptography is widely used on the Internet today. One example is the Data Encryption Standard (DES), which is not absolutely secure. Another is the *Vernam ciphers* (also called *one-time pad*), which are uncrackable but too slow for all but most sensitive transmissions. DES makes use a of 56-bit private key operating on 64-bit binary blocks. In Table 1.5 we give an example of a one-time pad cryptosystem using a smaller base-64 representation of basic ASCII characters.

Table 1.5. An example of *private key cryptography*: one-time pad.

1	original text	q	u	b	i	t
2	binary encoding (of 1)	101010	101110	011011	100010	101101
3	private key (control bits)	110010	010110	110110	111001	011010
4	XOR bits ($2 \oplus 3$)	011000	111000	101101	011011	110111
5	encrypted text (of 4)	Y	4	t	b	3
6	XOR bits ($4 \oplus 3$)	101010	101110	011011	100010	101101
7	decrypted text (of 6)	q	u	b	i	t

Let us look at the example shown in Table 1.5. Alice wants to encrypt a word of five characters and send it to Bob. Each character in base-64 is represented by a 6-bit string, and therefore any such word is represented by a 30-bit binary sequence. To encrypt it, Alice has to produce a key and a function that can encrypt the word using the key and decrypt it using the same key again. As we will see below, the best key is a completely random 30-bit binary sequence.[14] The appropriate function is XOR, $\oplus$ (see pp. 7, 8), since $(A \oplus B) \oplus B = A$. Next, she has to send the key to Bob via a very trusted carrier, if not in person. If she decided to send him the word *qubit*, she would obtain the encrypted message *Y4tb3* and could send it through any public channel to Bob, who can decrypt it using the key and the XOR function as shown in Table1.5.

The main advantage of this system is that it is unbreakable, provided that the key is used only once (hence the name *one-time* pad). It is unbreakable because the random key randomizes the encrypted text as well: namely, 0s and 1s in a binary representation of a character are unevenly distributed. Let the probability of having 0 in such representation be p. Then the probability of having 1 is $1 - p$. The probability of having either 0 or 1 in a random key is $1/2$. Hence, the probability of having 0 in the encrypted message after applying XOR is given by a sum of products of relevant probabilities: XOR gives 0 when we have either 0s or 1s in both the message and the key; the probability of the former case is $\frac{1}{2}p$ and of the latter $\frac{1}{2}(1 - p)$; their sum is $1/2$; and therefore, 0s appear in an encrypted message evenly and 1s equally so. However, if a key were used more than once, on two or more different messages, then one might be able to determine correlations between 0s and 1s in these encrypted messages and decipher the key.

The main disadvantages of this system are that Alice and Bob have to exchange the key through a reliable channel, they have to do so for each message anew, and the key has to be as long as the message. Public key cryptography solves the first two problems. The last problem is usually solved so as to leave the major part of communication nonencrypted and to encrypt just a signature or just the most sensitive data.

Classical public key cryptography relies on the assumed (sub)exponential complexity of factoring numbers. The most commonly used version RSA, was introduced in 1978 by Rivest, Shamir, and Adleman [Rivest et al., 1978] and is widely accepted today in spite of several shortcomings: a classical polynomially complex algorithm might be found; it can be up to 10^4 times slower than private cryptography, for example,

---

[14]If she wanted to have a genuinely random sequence, she should actually employ a proper quantum process already here, since classical processes or algorithms can provide her only with *pseudorandom* sequences.

DES; and it is software based, as opposed to DES, for which hardware has been developed.

The RSA public key protocol runs as follows:

1. Bob chooses two prime numbers p and q.

2. He calculates $n = pq$.

3. He selects e, which is *relatively prime* to $(p-1)(q-1)$ (two integers are relatively prime if their greatest common divisor is 1).

4. Pair $\{e, n\}$ is the *public key*. Bob sends it to Alice through a public channel.

5. Bob chooses an integer d that would reduce $(ed-1)/[(p-1)(q-1)]$ to an integer.

6. Pair $\{d, n\}$ is Bob's *private key*.

7. Alice uses public key $\{e, n\}$ to encrypt message m by means of the equation: $c = m^e \bmod n$, where *mod*ulo means the remainder after division. Then she sends c to Bob.

8. Bob decrypts the cyphertext c by means of the equation $m = c^d \bmod n$.

As an example, let us encrypt *qubit*. For convenience, we will switch to decimal representation. Using Eq. (1.66), we get $2^5+2^3+2 = 42$ for q, and 46, 27, 34, 45 for u, b, i, t, respectively. Thus, *qubit* is represented by 4246273445. In step one of the protocol, we choose (small) $p =$ prime[21] $= 73$ and $q =$ prime[18] $= 61$. We get $n = pq = 4453$ (step 2). For step 3 we find prime[606] $= 4457$ and prime[605] $= 4451$ and choose $e = 4457$. Finding a d that will reduce $r = (ed-1)/[(p-1)(q-1)]$ to an integer is not as straightforward, but a small program gives, for example, $d = 473$ and $r = 488$ and completes step 5. Now, because we have chosen a small n, we have to break 4246273445 into three parts and encrypt them piecewise. So step 7 gives $c = 4426^e \bmod n = 4034$, $c = 2734^e \bmod n = 3344$, and $c = 45^e \bmod n = 1513$, and the encrypted message reads: `oi hs PN`. Bob then uses step 8 to decrypt the message: $4034^d \bmod n = 4426 =$*qu*, $3334^d \bmod n = 2734 =$*bi*, $1513^d \bmod n = 45 =$*t*.

More realistic examples can be generated with the help of programs available on the web. They use hexadecimal numbers and extended ASCII (256) characters for encryption. In our case, 64-bit encryption suffices. We get e=`aa934cd8a567932b`, d=`1608a7af02c9c603`, $n =$ `c81f516f71fcb7c9`. This encrypts *qubit* as `1%^KimõE÷`. If we wanted to encrypt *qubit* so as to be unbreakable with today's technology for at least a few years—provided no one comes forward with a polynomial classical factoring algorithm in the meantime—we should use 1024-bit encryption. This gives us a key 8,500 hexadecimal digits long, which encrypts *qubit* in a string of over 300 extended ASCII characters. Generation of the keys, encryption, and decryption may take up to a few seconds on today's PCs, which is quite a long time for writing down just a few words[15] at a distance. A realistic application also requires a method of authenticating Alice, so she too has to produce her own public and private keys and combine them with Bob's.

What would solve all the problems with the protocols and algorithms we have presented here is a combination of the reliability of a one-time pad and a new, fast, and unbreakable way of exchanging keys. And this is exactly what the physics of quantum cryptography offers: not a quantum algorithm or "quantum software," but a solution based on the behavior of quantum systems themselves—quantum hardware.

From our examples of the roles of quantum entanglement and teleportation in quantum computation, we have learned that the possibility of using quantum systems for a new quantum technology depends on whether we can control the particular quantum feature we would like to use. In cryptography, the feature we would like to use is exactly what we already have with entanglement. On the one hand, there is a genuine randomness of measurable properties of subsystems that emerges from the complete absence of their predetermined properties, and on the other, there is the perfect correlation of the measurable properties of subsystems when we jointly carry out measurements on them. In other words, Alice communicates with Bob not by sending him a bit-by-bit message but by their joint recovery of bits from the correlation of their measurement clicks.

To see how this works, let us look at Fig. 1.27. Alice sends vertically, horizontally, and diagonally polarized photons (only the vertically polarized are shown in the figure) to Bob, who receives them through anisotropic birefringent plates that split incident beams into two beams with two different directions and polarizations. Beams exiting the plate are called *ordinary* and *extraordinary* rays. They are polarized at right angles to each other. In contrast to Sec. 1.8 (p. 17) and Fig. 1.6 (p. 17), where the polarizer absorbed perpendicularly oriented photons, a birefringent plate serving as a polarizing beam splitter always lets a photon through—either as an ordinary or as an extraordinary ray.

[15]More words than just *qubit*—up to 20 words as long as *qubit*—could be encrypted by one 1024-bit key.

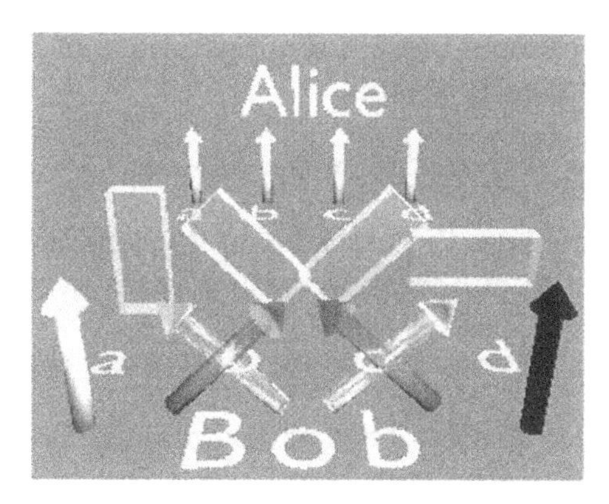

Figure 1.27. Physical scheme underlying quantum cryptography. Alice can send photons polarized in two bases: ⊞ and ⊠. Here, only vertically polarized photons, ↑ in basis ⊞ are shown. Photons pass through birefringent plates as either *ordinary* (bright arrows) or *extraordinary* (dark arrows) rays. Whether a photon moving along paths b and c will appear in an ordinary or an extraordinary ray is completely unpredictable.

In Fig. 1.27, we denote the polarization of the ordinary ray with a bright arrow and of the extraordinary ray with a dark arrow. Both the photons sent by Alice which are entering the plates, and the photons received by Bob which have exited the plates, can be oriented along four different directions: ↑, ↖, ↗, and →. In Fig. 1.27, however, only Alice's ↑ photons are shown. Of these vertically polarized photons sent by Alice along path a, Bob always receives only the ordinary ray polarized vertically and never along d. The chances of a photon appearing as either ↖ or ↗ from the diagonally oriented plates on paths b and c are 50:50. As it is obvious from Fig. 1.7 (p. 19) and Secs. 1.8 (p. 17) and 1.11 (p. 24), vectors ↑, → and ↖, ↗ can form two bases in either of which one can determine any polarization vector in our example. Let us denote these bases by ⊞ and ⊠, respectively. Of course, if Alice sent photons oriented along ↗, then Bob would always have clicks from ordinary photons at c, never from those at b, and on average every second time from ordinary photons at a and d. We denote qubits $|0\rangle$ and $|1\rangle$ in basis ⊞ as $|\uparrow\rangle$ and $|\rightarrow\rangle$, respectively, and in basis ⊠ as $|\nwarrow\rangle$ and $|\nearrow\rangle$.

There are several quantum cryptography protocols that use this method of communication. We first present BB84, named after Bennett and Brassard [Bennett and Brassard, 1984].

1. Alice chooses random data bits (0s and 1s).

2. She chooses bases ⊞ and ⊠ at random.

3. She sends qubits (photons) to Bob.

4. Bob randomly chooses bases, i.e., orientations of birefringent plates prior to receiving each photon.

5. He measures the polarization of the photons.

6. He publicly announces the bases he used whenever he detects a photon and Alice (also publicly) says which bases were correct.

7. They discard results corresponding to incorrect choices of bases.

8. To check whether *Eve* has been eavesdropping, Bob publicly reveals some of the results kept after step 7.

9. Alice confirms them. If they find that the results in step 8 differ unacceptably they abort the protocol.

10. If the results in step 8 do not differ significantly, the remaining bits are Alice and Bob's secret key.

We illustrate the protocol by the example shown in Table 1.6.

Table 1.6. An example of the BB84 protocol. The numbers in the first column correspond to the steps in the protocol. According to [Bennett and Brassard, 1984].

1	1	0	0	1	0	0	1	1	0	1	0	0	1	1	0
2	⊠	⊞	⊠	⊞	⊞	⊞	⊞	⊞	⊠	⊠	⊞	⊠	⊠	⊠	⊞
3	↗	↑	↖	→	↑	↑	→	→	↖	↗	↑	↖	↗	↗	↑
4	⊞	⊠	⊠	⊞	⊞	⊠	⊠	⊞	⊠	⊞	⊠	⊠	⊠	⊠	⊞
5	↖		↖		↑	↗	↗	→		↑	↑	↖		↗	↑
6	⊞		⊠		⊞	⊠	⊠	⊞		⊞	⊠	⊠		⊠	⊞
7			✓		✓			✓				✓		✓	✓
8					↑									↗	
9					✓									✓	
10			0					1				0			0

There are several points to be emphasized in the above protocol:

- Whenever Bob and Alice use the same bases, the vectors they obtain are (ideally) correlated, and whenever they use different ones, the vectors are uncorrelated and only 50% of obtained bits are correct. Therefore, Bob's error rate would have been 25%, if he had taken into account the results obtained in different bases. However, Bob and Alice discard the latter results altogether, i.e., they discard 50% (ideally) of all the results, following step 6, and the discarded 50%

include the aforementioned 25% errors.[16] The bits they keep we call the *sifted* key. Ideally, this key, would be flawless and unbreakable.

- In a realistic setup, Bob and Alice discard more than 50% of their results. This procedure compensates for one kind of error, such as poor single-photon detection. However, some other errors, like those stemming from imperfect alignment of Bob's vs. Alice's polarizers, cannot be directly detected since Bob and Alice cannot announce the results. For instance, when both Bob and Alice use basis ⊞, sometimes, when the alignment is not perfect, a photon polarized along ↑ emerges from Bob's polarizer as →. To correct such errors Bob and Alice can apply error correction schemes [see Sec. 1.22 (p. 72)] [Steane, 1996b; Calderbank and Shor, 1996] and entanglement purification (see p. 153) [Pan et al., 2001].
- Eve's eavesdropping will appear to Bob as a combination of both kinds of errors. Hence, he will be sure of her eavesdropping only if his error rate is high enough. Let us first consider the simplest eavesdropping method Eve could use: she puts a polarizer—randomly chosen to be oriented along either ↑ or → for ⊞ or along either ↖ or ↗ for ⊠—in the path of Alice's photons and keeps it there all the time. Then the information on the chosen bases that Alice and Bob exchange in step 6 of the protocol, together with the orientation of her polarizer, would give her 50% of Alice's bits when the orientation of her basis coincides with Alice's (Alice first throws away those bits that Bob does not have a record of, and thereupon Alice and Bob throw away bits for different bases.) Altogether, Eve can copy 25% of the results. As for Bob, he would not receive 50% of the photons at all, and 50% of the photons passing through Eve's incorrectly oriented polarizers would also be read off "incorrectly" by his birefringent plates. (When Alice uses ⊞ and ↑ and Eve ⊠, Bob gets ↑ photon through ⊞ in half of the measurements. So he recovers 50% of the photons that passed through Eve's incorrectly oriented polarizers.) So his error rate would be 62.5%, and the protocol is aborted. But Eve can decide to apply her strategy to only a fraction of Alice's bits, say 5%, and rely on getting more information when the key is applied to a message Alice would send to Bob later on. Then Eve's information is about 1.2% and Bob's error rate about 3.12%.

Taken together, the BB84 protocol appears to be reasonably robust and leaves intruders little chance. It is, however, physically interesting

[16]The point is the Bob cannot know which 25% are correct and which are wrong.

to see whether we can adapt this and possibly other protocols so as to cancel out an eavesdropper's attempts completely, i.e., whether quantum cryptography can be unconditionally secure.

1.22 Quantum Error Correction

To prove the unconditional security of quantum cryptography, it is, at least for the time being, enough to prove it for BB84. It therefore might seem surprising that we had to wait 15 years for the first such proofs [Mayers, 2001; Biham et al., 2000; Lo and Chau, 1999]. However, they all turn out to be reducible to particular forms of quantum error correction codes [Shor and Preskill, 2000; Calderbank and Shor, 1996; Steane, 1996b].

Quantum error correction theory adapts classical error correction theory to quantum states. Therefore, we shall briefly consider the classical theory first.

Error correction schemes are based on some kind of preparation (encoding) of our bits. One of the simplest such encodings is for "classical Alice" to add the so-called *parity bit* to all messages she sends to Bob. Here, *parity* is the quality of being odd or even. Alice and Bob agree that she should choose the parity bit 0 if the number of 1s in the message is even, and 1 if the number is odd. So if she wants to send 1001, she should add the parity bit 0 (because the number of 1s in 1001 is even) and send 10010 instead of 1001; if she wants to send 1101, she should add the parity bit 1 (because the number of 1s in 1101 is odd) and send 11011. If a *bit-flip* occurs and Bob receives, say 11111 in the last message, he would calculate $1 \oplus 1 \oplus 1 \oplus 1 = 0$ and would know that something went wrong because the parity bit 1 indicated that the parity of the message should have been odd.

We can formalize the procedure as follows. Alice encodes her messages by means of the following 4×5 matrix:

$$\begin{bmatrix} 1 & 0 & 0 & 1 \end{bmatrix} \begin{bmatrix} 1 & 0 & 0 & 0 & 1 \\ 0 & 1 & 0 & 0 & 1 \\ 0 & 0 & 1 & 0 & 1 \\ 0 & 0 & 0 & 1 & 1 \end{bmatrix} = \begin{bmatrix} 1 & 0 & 0 & 1 & 0 \end{bmatrix},$$

$$\begin{bmatrix} 1 & 1 & 0 & 1 \end{bmatrix} \begin{bmatrix} 1 & 0 & 0 & 0 & 1 \\ 0 & 1 & 0 & 0 & 1 \\ 0 & 0 & 1 & 0 & 1 \\ 0 & 0 & 0 & 1 & 1 \end{bmatrix} = \begin{bmatrix} 1 & 1 & 0 & 1 & 1 \end{bmatrix}. \tag{1.146}$$

Note that for calculating the last column we used XOR: $1 \oplus 1 \oplus 0 \oplus 1 = 1$. So, if there are no errors, Bob will always get 0 for the XOR value of encoded words: $1 \oplus 0 \oplus 0 \oplus 1 \oplus 0 = 0$, $1 \oplus 1 \oplus 0 \oplus 1 \oplus 1 = 0$. If there is an error, say 1111 as above, he will get $1 \oplus 1 \oplus 1 \oplus 1 \oplus 1 = 1$, which will detect it.

Of course, if two flips occur, Bob will not detect any error. Also, when Bob detects an error, he cannot tell which bit flipped. There are many other classical codes that can enable him to spot *and* correct the flipped bit. The first one is "brute force," i.e., just to repeat the same message three times in a row (two times would not suffice because Bob would not be able to not tell which one was right and which wrong). The information rate of brute force is very low: 1/3. There are, however, other codes that have a much higher information rate as well as better error detection and correction ability, which are the reasons error correction theories are so important. Here, we shall consider one of the latter codes, the *Hamming scheme*, which we will also use for a quantum error correction scheme.

We start with some definitions:

- ◇ A *codeword* is a word over {0,1}. The number of its bits is n.
- ◇ A *code* is a set C of codewords.
- ◇ An *error* is a change of bits on the way from Alice to Bob.
- ◇ *Data bits* make a message within a codeword Alice sends to Bob. Their number is d. We denote a codeword containing d data bits by (n,d).
- ◇ *Parity bits* are check bits within a codeword Alice sends to Bob. Parity bits enable Bob's error correction. Their number is p.
- ◇ The *Hamming distance between two words* is the number of bit positions in which the words disagree.
- ◇ The *Hamming distance of a code* C, denoted $D(C)$, is the minimum distance of two codewords in the code. A code (n,d) having a distance $D(C)$ will be denoted by $[n,d,D]$.
- ◇ A *Hamming code* is a code having $D(C) = 3$.
- ◇ The *information rate* of a code C of length n over alphabet F with size $|F| = q$ is $\log_q |C|/n$.

One can prove the following results [MacWilliams and Sloane, 1977]:

- The number of parity check bits required for each message is given by the following Hamming rule:

$$d + p + 1 \le 2^p. \qquad (1.147)$$

- A code with distance D is (D − 1) error detecting and (D − 1)/2 error correcting. Hence, the Hamming code allows one to detect and correct one error.

- For D = 3, only the following (n, d) codes are possible:

$$(n, d) = (2^i - 1, 2^i - 1 - i),$$

where $i \geq 3$ is an integer. So, for instance, we can have the following Hamming codes: (7,4), (15,11), (31,26), etc.

- An (n, d) code is a d-dimensional subspace of F^n, whose size is $q = |F|$. Since $|C| = q^d$, the information rate of the code is $\log_q |C|/n = d/n$.

To see how Bob can correct an error using the Hamming scheme, let us consider the Hamming code (7,4), for which the information rate is $d/n = 4/7 \approx 0.57$, i.e., about 1.7 times higher than the previously mentioned triple sending. It follows from Eq. (1.147) that $p = 3$. One can easily check that among $2^7 = 128$ possible 7-bit codewords, there are only 16 valid ones (with D=3). These can be generated by, for example, the following *generation matrix*:

$$G = \begin{bmatrix} 1 & 1 & 1 & 1 & 1 & 1 & 1 \\ 0 & 0 & 0 & 1 & 1 & 1 & 1 \\ 0 & 1 & 1 & 0 & 0 & 1 & 1 \\ 1 & 0 & 1 & 0 & 1 & 0 & 1 \end{bmatrix}, \tag{1.148}$$

with the help of which we encode message m so as to get the codeword $c = m \cdot G$. Note that, in multiplication of matrices we make use of XOR, i.e., of adding their elements modulo 2; the dot "·" refers to this. For 4-bit messages, Alice gets the following codewords:

Table 1.7. Unique Hamming codewords $c = m \cdot G$ (top lines) for messages m (bottom lines) that Alice sends to Bob.

0000	0001	0010	0011	0100	0101	0110	0111
0000000	1010101	0110011	1100110	0001111	1011010	0111100	1101001

1000	1001	1010	1011	1100	1101	1110	1111
1111111	0101010	1001100	0011001	1110000	0100101	1000011	0010110

Bob receives the codeword c, and using the *check matrix* H, defined below in Eq. (1.149), he gets $H \cdot c^T = s$. In the case of no errors for the above codeword, he will always get $s = 0$, as one can easily check, and then he recovers the original message according to Table 1.7. If, for example, the codeword 0011001 had the sixth bit flipped to 1 and Bob received the string 0011011, he would obtain

$$H \cdot c^T = \begin{bmatrix} 0 & 0 & 0 & 1 & 1 & 1 & 1 \\ 0 & 1 & 1 & 0 & 0 & 1 & 1 \\ 1 & 0 & 1 & 0 & 1 & 0 & 1 \end{bmatrix} \cdot \begin{bmatrix} 0 \\ 0 \\ 1 \\ 1 \\ 0 \\ 1 \\ 1 \end{bmatrix} = s = \begin{bmatrix} 1 \\ 1 \\ 0 \end{bmatrix}. \qquad (1.149)$$

This s is called the *syndrome*, and it tells us which column we should look at in H. In this case it is the sixth column, meaning that the sixth bit in the string has been flipped. So, after correcting it, we get the original codeword 0011001. Note that H is a submatrix of G: it equals the second through fourth rows of G. Also note that the codeword 0011001 is a unique codeword with distance 1 from string 0011011. This is why we need codewords with distances of at least 3 from each other. The following general results hold:

- For one error in the ith bit, the syndrome s is the ith column of H.
- Each error of *weight* (number of erroneous bits) up to $(\mathrm{D}-1)/2$ has a unique syndrome.

We can look at the error correction in the following way. The codeword u Alice sent is affected by noise in the communication channel so that it changes into $u' = u + e$, where e is the error caused by the noise and where the addition is modulo 2. In the above example, $u = 0011001$, $u' = 0011011$, and $e = 0000010$. From u' Bob can uniquely recover u using the code C, but he needs not learn how the word actually reads because

$$H \cdot u' = H \cdot (u + e) = H \cdot u + H \cdot e = H \cdot e = s, \qquad (1.150)$$

and this means that he can learn the syndrome without ever learning the word. (For the code [7,4,3] there are 16 codewords and only 7 syndromes.) Of course, in the classical case he can always look at Table 1.7, but in the quantum case, this amounts to correcting a quantum state without disturbing it, and this outcome is what we are looking for in quantum error correction and quantum cryptography [Steane, 1998].

While classical error correction protocols encode bits that Alice transmits to Bob by means of additional parity bits, quantum protocols must be able to encode superposed qubits by entangling them with additional qubits because we cannot make more replicas of a superposed state (see the no-cloning theorem, p. 58). Superpositions are what we essentially have both in quantum computation and in quantum cryptography.

The errors that can occur in transmission (in quantum cryptography as well as in quantum computation) of an arbitrary qubit state

$$|\psi\rangle = \alpha|0\rangle + \beta|1\rangle, \qquad \text{where } |\alpha|^2 + |\beta|^2 = 1, \tag{1.151}$$

are a *bit-flip*,

$$\mathrm{X}|\psi\rangle = \begin{bmatrix} 0 & 1 \\ 1 & 0 \end{bmatrix} |\psi\rangle = \beta|0\rangle + \alpha|1\rangle, \tag{1.152}$$

a *phase shift*,

$$\mathrm{Z}|\psi\rangle = \begin{bmatrix} 1 & 0 \\ 0 & -1 \end{bmatrix} |\psi\rangle = \alpha|0\rangle - \beta|1\rangle, \tag{1.153}$$

or both:

$$\mathrm{Y}|\psi\rangle = \mathrm{ZX}|\psi\rangle = \begin{bmatrix} 0 & 1 \\ -1 & 0 \end{bmatrix} |\psi\rangle = \beta|0\rangle - \alpha|1\rangle, \tag{1.154}$$

where $\mathrm{X} = \hat{\sigma}_x$, $\mathrm{Z} = \hat{\sigma}_z$, and $\mathrm{Y} = i\hat{\sigma}_x$ (cf. Eq. (1.88)).

Let us first consider bit-flip correction. If we encode a single qubit in the state $\psi = \alpha|0\rangle + \beta|1\rangle$ by means of entangled qubits whose sequence corresponds to classical words, we will be able to use classical error correction applied not to the original superposition but to a superposition of such quantum codewords. And this is exactly what Steane did with the Hamming code [Steane, 1996a]. A similar approach was taken by Calderbank and Shor [Calderbank and Shor, 1996]. Hence the name *CSS codes.*

The idea put forward in [Steane, 1996a; Steane, 1996b] was to generate Hamming codewords using quantum gates so as to enable error detection and correction analogous to that given by Eq. (1.149). The encoding is shown in Fig. 1.28 below.

In this approach we do not have four data bits that we encode by means of additional three bits, as in the classical Hamming code, but we will still be able to use this classical code for the error correction. Instead of four bits, we encode just one qubit in an unknown state by entangling it with with six other qubits, all initially in state $|0\rangle$. The qubit we encode is $|\psi\rangle$ in Fig. 1.28, given by Eq. (1.151). As in Fig. 1.24, when we apply the control qubit $|\psi\rangle$ (third qubit) to the fifth and sixth qubits ($|0\rangle_5$ and $|0\rangle_6$) as its targets, we entangle them:

$$|\psi_{356}\rangle = \alpha|0\rangle_3|0\rangle_5|0\rangle_6 + \beta|1\rangle_3|1\rangle_5|1\rangle_6, \tag{1.155}$$

Consequently, from this point on, the lines that originally represented the third, fifth, and sixth qubit now represent one and the same state.

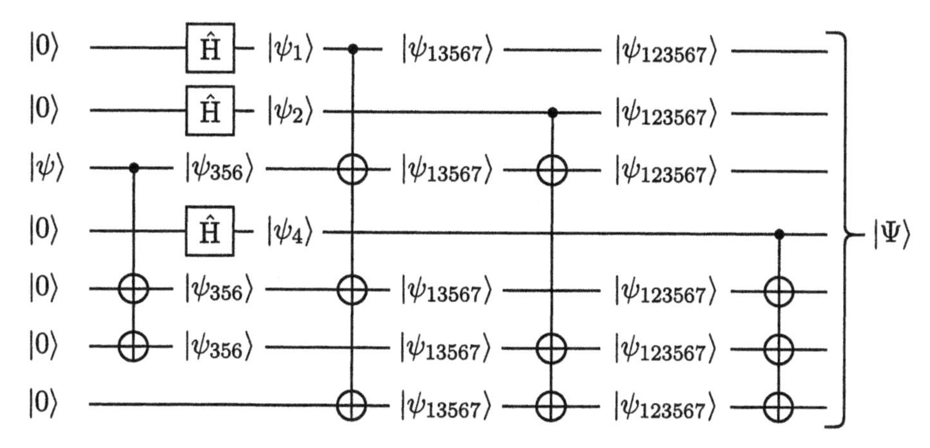

Figure 1.28. A 7-qubit encoding of $|\psi\rangle = \alpha|0\rangle + \beta|1\rangle$ into $|\Psi\rangle = \alpha|0\rangle_{1-7} + \beta|1\rangle_{1-7}$ according to [Steane, 1996b]. H is the Hadamard matrix given by Eq. (1.59).

The next time point corresponds to $|\psi_1\rangle = |0\rangle_1 + |1\rangle_1$ acting as a control qubit on the third, fifth, and sixth qubits. The third and fifth qubits are entangled with the sixth qubit, so it is brought in as well. The resulting state is an entanglement of all five qubits:

$$\begin{aligned}|\psi_{13567}\rangle &= \alpha|0\rangle_1|0\rangle_3|0\rangle_5|0\rangle_6|0\rangle_7 + \beta|0\rangle_1|1\rangle_3|1\rangle_5|1\rangle_6|0\rangle_7 \\ &\quad +\alpha|1\rangle_1|1\rangle_3|1\rangle_5|0\rangle_6|1\rangle_7 + \beta|1\rangle_1|0\rangle_3|0\rangle_5|1\rangle_6|1\rangle_7. \end{aligned} \tag{1.156}$$

At the next level, $|\psi_2\rangle = |0\rangle_2 + |1\rangle_2$ entangles all but the fourth qubit:

$$\begin{aligned}|\psi_{123567}\rangle &= \alpha|0\rangle_1|0\rangle_2|0\rangle_3|0\rangle_5|0\rangle_6|0\rangle_7 + \beta|0\rangle_1|0\rangle_2|1\rangle_3|1\rangle_5|1\rangle_6|0\rangle_7 \\ &\quad +\alpha|1\rangle_1|0\rangle_2|1\rangle_3|1\rangle_5|0\rangle_6|1\rangle_7 + \beta|1\rangle_1|0\rangle_2|0\rangle_3|0\rangle_5|1\rangle_6|1\rangle_7 \\ &\quad +\alpha|0\rangle_1|1\rangle_2|1\rangle_3|0\rangle_5|1\rangle_6|1\rangle_7 + \beta|0\rangle_1|1\rangle_2|0\rangle_3|1\rangle_5|0\rangle_6|1\rangle_7 \\ &\quad +\alpha|1\rangle_1|1\rangle_2|0\rangle_3|1\rangle_5|1\rangle_6|0\rangle_7 + \beta|1\rangle_1|1\rangle_2|1\rangle_3|0\rangle_5|0\rangle_6|0\rangle_7. \end{aligned} \tag{1.157}$$

In the end, $|\psi_4\rangle = |0\rangle_4 + |1\rangle_4$ gives

$$\begin{aligned}|\Psi\rangle &= \alpha(|0000000\rangle + |1010101\rangle + |0110011\rangle + |1100110\rangle \\ &\qquad +|0001111\rangle + |1011010\rangle + |0111100\rangle + |1101001\rangle) \\ &\quad +\beta(|1111111\rangle + |0101010\rangle + |1001100\rangle + |0011001\rangle \\ &\qquad +|1110000\rangle + |0100101\rangle + |1000011\rangle + |0010110\rangle) \\ &= \alpha|0\rangle_{1-7} + \beta|1\rangle_{1-7}, \end{aligned} \tag{1.158}$$

and these are nothing but the Hamming codewords from Table 1.7 (p. 1.7).

Hence, for correcting errors we can use classical Hamming theory if we assume that we will mostly have only one-flip errors, i.e., that the probability of having two such errors within a transmission of one codeword is negligible. The check matrix H given by Eq. (1.149) then tells us how to design the error-correcting scheme. In the matrix multiplication of codewords by H modulo 2, 1s in H turn 1s in a codeword into 0 and 0 into 1. In other words, they behave like control qubits. Recalling Eq. (1.150), we see that their targets can be three syndrome qubits $|0\rangle$, i.e., three additional check qubits—usually called *ancillas.* In this way we get around the lack of four data bits we would have in the classical Hamming code. How this can be carried out is shown in Fig. 1.29.

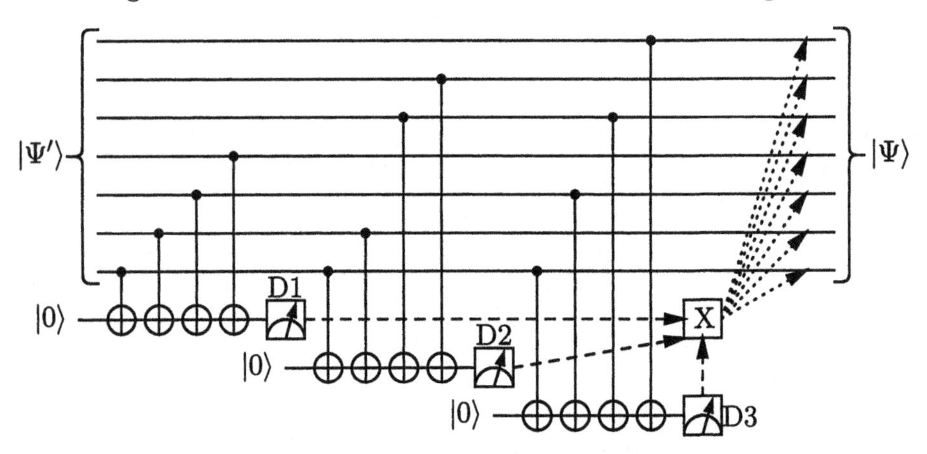

Figure 1.29. A 7-qubit error-correcting code. The control qubits are chosen so as to correspond to 1s in H, as given by Eq. (1.149). Detectors D1–3 measure the ancillas and apply flip-correcting X to a qubit found to have suffered a bit-flip in transmission. The input state $|\Psi'\rangle$ is $|\Psi\rangle$ (from Fig. 1.28) with possibly a flipped qubit state.

The rows in H given by Eq. (1.149) determine the control qubits acting on each of three ancillas in Fig. 1.29. Checking all terms in Eq. (1.158), we see that the parity of all bits corresponding to the first, second, and third row of H is always even, and when an even number of flips occurs, each ancilla $|0\rangle$ will be flipped twice and detectors D1–3 will find them in the state $|0\rangle$. If an odd number of flips—i.e., an error in transmission—occurs, one, two, or all three detectors will detect the state $|1\rangle$. According to the correspondence with the columns in H, the flipped qubit state will be corrected by means of X. For example, if D1 and D2 detected the state $|1\rangle$, the state of sixth qubit will be flipped back by means of X because $[110]^T$ is the sixth column in H (Eq. (1.149)).

After the correction has been carried out and the codestate $|\Psi\rangle$ restored, Bob has to decode $|\Psi\rangle$ so as to obtain $|\psi\rangle$ sent by Alice. (We stress again here that this process is the same whether Alice and Bob are simply parts or stages of quantum computation or parties in quantum communication.) Bob decodes the message by reversing the procedure given in Fig. 1.28 (p. 77) while substituting Z for H. To clarify the reversed procedure, let us just consider its last step, restoring $|\psi\rangle$ from $|\psi_{356}\rangle$, as given in Fig. 1.30.

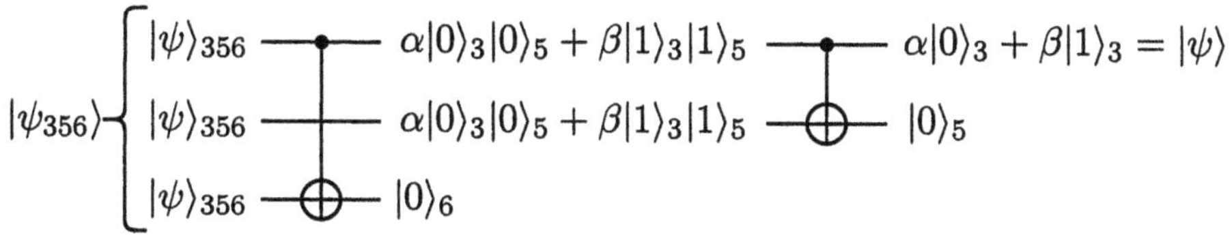

$|\psi_{356}\rangle$: $|\psi\rangle_{356}$ — $\alpha|0\rangle_3|0\rangle_5 + \beta|1\rangle_3|1\rangle_5$ — $\alpha|0\rangle_3 + \beta|1\rangle_3 = |\psi\rangle$

$|\psi\rangle_{356}$ — $\alpha|0\rangle_3|0\rangle_5 + \beta|1\rangle_3|1\rangle_5$ — $|0\rangle_5$

$|\psi\rangle_{356}$ — $|0\rangle_6$

Figure 1.30. The last two levels of decoding a superposition of 7-qubit codewords $|\Psi\rangle$ encoded in Fig. 1.28 and corrected in Fig. 1.29.

We can say that the state $|\psi_{356}\rangle$ given by Eq. (1.155) "acts on itself" in the following sense. First, within each product state (codeword) of the state $|\psi_{356}\rangle$, the third qubit state acts as a control qubit on the sixth qubit state. Hence, $|0\rangle_3$ from $|0\rangle_3|0\rangle_5|0\rangle_6$ acts on $|0\rangle_6$ from this product and leaves it unchanged, while $|1\rangle_3$ from $|1\rangle_3|1\rangle_5|1\rangle_6$ flips $|1\rangle_6$ into $|0\rangle_6$. As a consequence, the sixth qubit disentangles from the third and fifth. In the next step, the fifth qubit disentangles from the 3rd one and we recover the original state $|\psi\rangle$of the latter qubit.

To see how we can correct the phase shift in Steane's 7-qubit code, let us start with Eq. (1.158):

$$|\Psi\rangle = \alpha|0\rangle_{1-7} + \beta|1\rangle_{1-7}. \tag{1.159}$$

Notice that by comparing terms in Eq. (1.158), we can see that $|1111111\rangle = \overline{\mathrm{X}}|0000000\rangle$, $|0101010\rangle = \overline{\mathrm{X}}|1010101\rangle$, etc., where $\overline{\mathrm{X}} = \mathrm{X} \otimes \cdots \otimes \mathrm{X}$, where $X = \mathrm{NOT} = \sigma_x$. Therefore, $|1\rangle_{1-7} = \overline{\mathrm{X}}|0\rangle_{1-7}$.

Using the Hadamard gate (p. 32), we can introduce the following new basis for each qubit:

$$|\overline{0}\rangle_i = \mathrm{H}|0\rangle_i = \frac{1}{\sqrt{2}}(|0\rangle_i + |1\rangle_i)$$
$$|\overline{1}\rangle_i = \mathrm{H}|1\rangle_i = \frac{1}{\sqrt{2}}(|0\rangle_i - |1\rangle_i), \qquad i = 1, \ldots, 7, \tag{1.160}$$

where we stripped the hat "ˆ" from H to ease the notation.

An attractive feature of our code is that upon substituting $|0\rangle_i = (|\overline{0}\rangle_i + |\overline{1}\rangle_i)/\sqrt{2}$ and $|1\rangle_i = (|\overline{0}\rangle_i - |\overline{1}\rangle_i)/\sqrt{2}$, $i = 1, \ldots, 7$ into Eq. (1.158), i.e., into Eq. (1.159), we get (the process is straightforward but tedious):

$$|0\rangle_{1-7} = \frac{1}{\sqrt{2}}(|\overline{0}\rangle_{1-7} + |\overline{1}\rangle_{1-7})$$
$$|1\rangle_{1-7} = \frac{1}{\sqrt{2}}(|\overline{0}\rangle_{1-7} - |\overline{1}\rangle_{1-7}). \tag{1.161}$$

This outcome means that the new basis contains only those states that are in the Hamming code—actually the same states that the old basis contains. In the new basis, $|\Psi\rangle$ reads

$$|\Psi\rangle = \frac{1}{\sqrt{2}}[(\alpha+\beta)|\overline{0}\rangle_{1-7} + (\alpha-\beta)|\overline{1}\rangle_{1-7}]. \tag{1.162}$$

Now, a phase shift in the old basis

$$|\Psi\rangle = \alpha|0\rangle_{1-7} - \beta|1\rangle_{1-7} \tag{1.163}$$

transforms in the new basis to

$$|\Psi\rangle = \frac{1}{\sqrt{2}}[(\alpha-\beta)|\overline{0}\rangle_{1-7} + (\alpha+\beta)|\overline{1}\rangle_{1-7}], \tag{1.164}$$

which is therefore nothing but a bit-flip (by definition—cf. Eq. (1.152)) in the latter basis.

Alternatively, we can use

$$\alpha|0\rangle_{1-7} - \beta|1\rangle_{1-7} = \overline{\mathrm{Z}}\,|\Psi\rangle = \overline{\mathrm{H}}\,\overline{\mathrm{X}}\,\overline{\mathrm{H}}\,|\Psi\rangle, \tag{1.165}$$

where $\overline{\mathrm{H}} = \mathrm{H}\otimes\cdots\otimes\mathrm{H}$ and $\overline{\mathrm{Z}} = \mathrm{Z}\otimes\cdots\otimes\mathrm{Z}$, since we can show that the following transformations hold:

$$\overline{\mathrm{H}}\,|0\rangle_{1-7} = \frac{1}{\sqrt{2}}(|0\rangle_{1-7} + |1\rangle_{1-7}), \quad \overline{\mathrm{H}}\,|1\rangle_{1-7} = \frac{1}{\sqrt{2}}(|0\rangle_{1-7} - |1\rangle_{1-7}).$$

Eqs. (1.162) and (1.165) mean not only that we can reduce the phase shift correction to a bit-flip correction with the same error correcting code but also that we can completely separate the corrections and do them in sequence, as shown in Fig. 1.31.

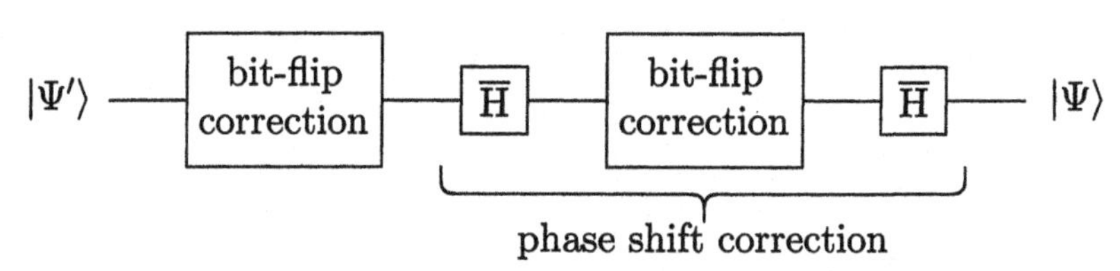

Figure 1.31. The 7-qubit correction code carries out bit-flip and phase-shift corrections with the same correction circuit (Fig. 1.29) in two bases.

The error correction scheme presented above enables us not only to correct errors in transmission but also to correct malfunctioning of the gates (CNOT) used to encode the message (cf. Fig. 1.28, p. 77). Such quantum computation, where we can correct malfunctioning of circuits, is called *fault-tolerant* computation [Steane, 1998].

For quantum computation and transmission with more than one error per code, classical codes with larger Hamming distances capable of correcting multiple errors [MacWilliams and Sloane, 1977] can be adapted in a way similar to the one presented above.

1.23 Unconditional Security of Quantum Cryptography

Before we dwell on the actual proof of the unconditional security, we will present variations of BB84 with respect to number of required states, realistic robustness, and underlying physical schemes.

Quantum cryptography protocol BB84, presented in the previous section, makes use of basically one single quantum feature: individual system states prepared in one basis (say ⊞) might be totally uncorrelated with individual states of the same system measured in some other basis (say ⊠). This feature suffices for secure distribution of secret keys, which is essentially what quantum cryptography is all about: a secure replacement for insecure classical public key distribution. All other parts of cryptocommunication remain classical. This is why it is often stressed in the literature that quantum cryptography should actually be called *quantum key distribution* [Gisin et al., 2002].

There are actually many varieties of quantum key distribution protocols [Gisin et al., 2002]. First, we need not keep to four states, as in BB84. Two states are enough, as the so-called B92 demonstrates [Bennett, 1992], but the security of B92 is lower than the security of BB84. A six-state protocol [Bruss, 1998] reduces Eve's information gain for a given error rate [Gisin et al., 2002] but is more demanding. Hence, BB84 tends to be standard.

Next, polarization is very suitable for understanding and carrying experiments in a laboratory, but it is not robust enough to allow implementation over larger distances (more than a few kilometers).

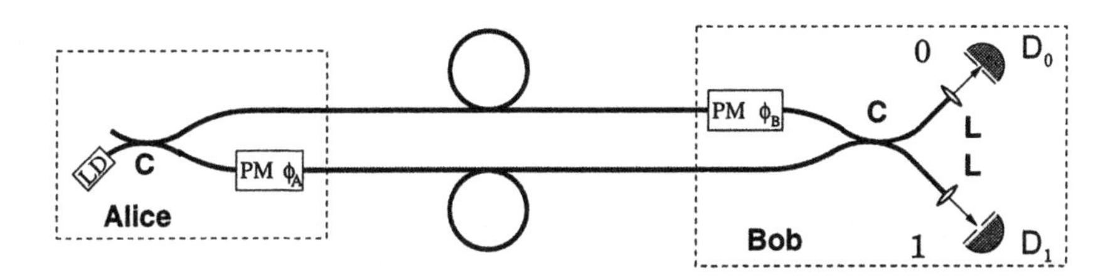

Figure 1.32. Phase-coding scheme of quantum cryptography: optical fiber Mach–Zehnder interferometer. LD is a laser diode; PM, phase modulators; C, symmetric fiber couplers—equivalent to beam splitters; D, avalanche detectors; L, lenses.

Fig. 1.32 shows a phase-coding scheme that is nothing but an optical fiber version of the Mach–Zehnder interferometer shown in Fig. 1.10. Here, beam splitters are substituted by fiber couplers (optical devices that merge two fibers). The probabilities of detectors D_0, D_1 registering a photon are given by Eq. (1.62):

$$p_0 = \cos^2 \frac{\phi_A - \phi_B}{2}, \qquad p_1 = 1 - p_0 = \sin^2 \frac{\phi_A - \phi_B}{2}. \tag{1.166}$$

Alice makes use of four phase shifts $\phi_A = 0, \pi/2, \pi, 3\pi/2$ and associates bit 0 with $\phi_A = 0$ and $\phi_A = \pi/2$ and bit 1 with $\phi_A = \pi$ and $\phi_A = 3\pi/2$. Bob makes use of two phase shifts $\phi_B = 0, \pi/2$ and associates bit 0 with a click of detector D_0, i.e., with $p_0 = 1$, and bit 1 with a click of detector D_1, i.e., with $p_1 = 1$. They have to discard cases when there is a 50:50 probability of either D_0 or D_1 clicking, i.e., when $p_0 = p_1 = 1/2$. This happens, for instance, when Alice chooses $\phi_A = 0$ and Bob $\phi_B = \pi/2$ since then $p_0 = p_1 = \cos^2(-\pi/4) = 1/2$. Hence they can implement the BB84 protocol as shown in Table 1.8 following [Gisin et al., 2002].

Table 1.8. Phase-coding implementation of BB84 protocol [Gisin et al., 2002].

Alice's bits	ϕ_A	ϕ_B	p_0	p_1	Bob's bits
0	0	0	1	0	0
0	0	$\pi/2$	1/2	1/2	undetermined
0	$\pi/2$	0	1/2	1/2	undetermined
0	$\pi/2$	$\pi/2$	1	0	0
1	π	0	0	1	1
1	π	$\pi/2$	1/2	1/2	undetermined
1	$3\pi/2$	0	1/2	1/2	undetermined
1	$3\pi/2$	$\pi/2$	0	1	1

In realistic applications, it is difficult to control the lengths of the two fibers in the above setup up to a fraction of the wavelength of photons. Specifically, for unequal paths, Eq. (1.166) reads

$$p_0 = \cos^2 \frac{\phi_A - \phi_B + k\Delta L}{2}, \tag{1.167}$$

where k is the wave number and ΔL is the path-length difference.

Therefore, variations on phase-coding that use only one fiber, with pulses going through it with a delay, have been put forward as most suitable for a practical implementation. For instance, both Alice and Bob can have their own Mach–Zehnder interferometers with unequal paths at each side, as proposed by Charles Bennett [Bennett, 1992]. A photon taking the shorter path in Alice's interferometer and the longer in Bob's cannot be distinguished from a photon taking the longer path in Alice's interferometer and the shorter in Bob's, and so we obtain the desired interference. Successful experiments have been carried out with fibers from 10 km [Townsend et al., 1993] to 48 km long [Hughes et al., 2000]. An even more robust design with two photon pulses traveling from Bob to Alice and back to Bob over the same fiber has been implemented [Gisin et al., 2002] and called *plug-and-play quantum cryptography*, since it requires no adjustment prior to usage.

As for the physical schemes underlying possible quantum key distribution protocols, apart from the single photons of BB84, we can also use a pair of photons entangled within an EPR pair, as given by Eq. (1.77). That is, instead of the scheme shown in Figs. 1.6 (p. 18) and 1.27 (p. 69), we can use the scheme shown in Fig. 1.19 (p. 44). However, by comparing the probability function of single polarized photons[17] with the probability function for two photons entangled in polarization[18], we see that the two schemes are completely equivalent. Thus, we may dwell on the proof of the unconditional security of BB84, which amounts to such proof for quantum cryptography in general.

As mentioned above, all previous proofs of the unconditional security of quantum cryptography [Mayers, 2001; Biham et al., 2000; Lo and Chau, 1999] were reduced by Shor and Preskill [Shor and Preskill, 2000] to a proof for BB84 based on quantum error correction and privacy amplification. The idea behind this reduction is that errors in transmission caused by technical imperfections are indistinguishable from the effects of Eve's eavesdropping. So, the proof of unconditional security of quantum cryptography consists of showing that by error correction we can reduce both the difference in Alice's and Bob's keys and the percentage of the key Eve can possess to arbitrarily small amounts.

In general, to enable Alice and Bob to exchange a key *unconditionally securely* we should, in addition to the BB84 protocol from Sec. 1.21 (p. 64), also assume that they have a quantum computer at their disposal to store qubits in its memory and to correct bit-flips and phase shift errors in transmission of qubits. Fortunately, for an error correction scheme that has bit-flip and phase shift correction completely separated, as for example the CSS scheme presented in Sec. 1.22 (p. 72), we can dispense with the quantum computer because we would only need it to carry out an interwoven bit-flip and phase shift error correction. Using a CSS scheme, however, we can keep the bit-flip correction and average

[17] The probability of photons coming out from the first polarizer being detected after passing through the second one—given by Eq. (1.9) for the first scheme.

[18] The probability of photons being detected by the two detectors shown in Fig. 1.19 (p. 44)—given, for example, by Eq. (1.143) for the second scheme.

over random phase vectors obtained by the syndrome measurements and get a mixed state that is equivalent to a randomly chosen code string [Shor and Preskill, 2000]. A bit-flip correction suffices for a complete recovery of all bits changed in transmission.[19]

Taken together, in addition to the points of the BB84 protocol from Sec. 1.21 (p. 64) we should, according to Shore and Preskill [Shor and Preskill, 2000], further assume the following:

11. Alice publicly announces $u+v$ where v is the sifted key block (the remaining bits they agreed upon in point 10) and u is a random codeword in Steane's code C. Bob adds[20] $u+v$ to his string $v+e$ and corrects the result $u+e$ he received through the quantum channel to the codeword in C.
12. Alice and Bob use the coset $u+C'$, $C' \subset C$ (see below) as their key.

This points deserves some comments: (a) The error correction scheme is reduced to the classical case. For the sake of simplicity we will use the simplest Hamming code C, although there are other more sophisticated and bigger codes, in particular those that can correct more than one error. (b) The length of the block Alice chooses to send must match the length of the codeword. In our case, both v and u contain seven bits. (c) v might but need not belong to the code C, i.e., Bob cannot correct v directly. (d) Eve cannot make use of $u+v$ to increase her information because Alice picks up bits for v at random, and the probability that Eve already has all these bits is negligible. So Eve will have more than one error and cannot correct these errors.

Let us consider the following example for these points with Alice picking up seven measurements among all those she and Bob sifted in step 10 of the BB84 protocol in Sec. 1.21. She knows she sent, for example, $v = [1110001]$. Bob, however, might have received it with a bit-flip: $v+e = [1111001]$. Neither v nor $v+e$ are Hamming codewords. Alice randomly picks up a codeword, for example, $u = [0011001]$, and publicly announces $u+v = [1101000]$. Bob adds it to $v+e$ and gets $v+e+u+v = [0010001] = u+e$. By using the check matrix for the Hamming code given by Eq. (1.149), Bob gets the syndrome $s^T = H \cdot [0010001]^T = [100]^T$. This syndrome s points to the fourth column of h, and he learns that his fourth bit has flipped.

---

[19] Here, as in Sec. 1.22 (p. 72), we assume that there is, on average, one error per message block within a key.

[20] In the literature one often finds the term "subtracts" here. But for XOR these two operations reduce to each other.

Now we turn to the code C' and the *coset* of C'—but only to go around it below. C' must be such that its dual code $C'^{\perp}$ has the same minimal distance (in our case, 3) as C [Steane, 1996b]. The coset of C' determined by u is the set of all the words of the form $u + c$ as c ranges over all words in C': $u + C' = \{u + c | c \in C'\}$. The generator G' of the dual code $C'^{\perp}$ gives the word $G'(u)$, which is in one-to-one correspondence with the coset of C': $G'(u) \to u + C$ [Lo et al., 2005]. Instead of giving the details of computing this one-one correspondence we would rather present a shortcut that amounts to the same result.

Let us substitute the following step for the step 12 above:

$\widehat{12}$. Bob uses the obtained e to correct his received string $v+e+e = v$, and thus Alice and Bob share the identical key-part v.

After Alice and Bob repeat steps 11 and $\widehat{12}$ enough times, they become almost certain that they have identical copies of the key.

However, computing the coset in step 12 would provide Alice and Bob with *privacy amplification*, and in using $\widehat{12}$ they have to do it separately. This is necessary because Eve can still be in possession of some parts of the key corresponding to the photons she measured and resent to Bob in a correct basis. To reduce the information that Eve can possess, Alice and Bob can apply the following amplification procedure. They pick a pair of bits (informing each other of their choice through a public channel) and substitute it with an XORed value of the pair. Since there is a high probability that Eve will not have information on both bits, this reduces her knowledge. Alice and Bob can iterate the procedure until the information Eve has gained cannot jeopardize the security of their key anymore.

This completes our sketch of the proof of the unconditional security of quantum cryptography. The maximum tolerable error rate is computed to be 11% [Shor and Preskill, 2000; Lo et al., 2005], but we have not elaborated on error and reliability estimates, since too many experimental improvements have been achieved recently. For example, last year, the experimentally achieved distances of transmission exceeded the ones of only five years ago by 300% [Corndorf et al., 2004]. In Secs. 2.7 (p. 125) and 3.1.7 (p. 151) we give an overview of recent experimental implementations and perspectives.

Chapter 2

EXPERIMENTS

2.1 Technological Candidates for Quantum Computers

To date—about ten years after the first experimental implementation of one qubit—most of the numerous proposals for quantum computing prototypes have not been able to implement more than one or two qubits. Thus, one still cannot single out a most promising technological candidate for a future quantum computer. This situation does not mean, however, that there are doubts about whether to pursue the experimental efforts further. The quantum computer project is taken as inevitable, and the experimental efforts have been undertaken and supported with a vigor similar to the efforts dedicated to the nuclear fusion program from the 1960s till today.

Both these projects appear to be unstoppable multi-billion dollar international initiatives (in 2004 the United States dropped its own fusion research project, FIRE, in favor of concentrating all resources on the international ITER project), but the quantum information processing project may have a better chance of success. There are several reasons for this. The nuclear fusion project is not perceived as being able to lead to a practical commercial power plant in the near future, even if the project turns out to be successful in the end, while an operating quantum computer would have no such problems, however complicated and expensive it would be. Moreover, the main classical computer producers increasingly support both theoretical and experimental quantum information groups. From the very beginning, nuclear fusion experiments and their financial support were mostly concentrated on the *tokamak*. In contrast, there is an increasing variety of comparatively low-cost platforms for quantum computers, and the financial support is spread evenly over a constantly growing number of experimental and theoretical groups throughout the world. Finally, nuclear fusion theory is basically a nuclear reaction theory tied mainly to the tokamak experiments, while quantum information theory is a true interdisciplinary project involving many specialties of physics, mathematics, and computer science. As a result, new jobs in quantum information field become available practically each day, and both quantum information theory and its experiments are having a growing impact on all the various fields from which they emerged.

At the beginning of this section, we mentioned that only a few technological candidates for future quantum computing platforms have succeeded in manipulating several qubits. These are nuclear magnetic resonance (NMR), with molecules in a liquid; ion traps; and cavity quantum electrodynamics (QED). In this chapter, we shall elaborate on the first two as well as a recently proposed silicon-based model. Then, in Sections 2.6 (p. 123) and 2.7 (p. 125), we shall briefly present a *quantum information science and technology roadmap* for quantum computation and communication, as well as some experimental details and perspectives of quantum cryptography setups.

2.2 Zeeman Effects

Qubits that we can prepare, manipulate, and measure are two-level states of photons, electrons, and nucleons. Apart from all-photon computers, electrons play an important role in understanding and designing quantum computers. In this section we therefore consider electron states. There are no "off the shelf" two-level states of electrons within atoms, since electron spins interact with their orbits and with the nuclei of the atoms. We can distinguish all the electron states only when atoms are in a magnetic field. This behavior is called the normal and anomalous electron Zeeman effect within atoms [Greiner, 1989]. We will review the Zeeman effects in this section, not because they work directly as quantum computing devices but because many details of quantum devices that we will elaborate on in the subsequent sections are based on or derived from them.

Let us consider an electron in the simplest possible atom (hydrogen) in the Bohr model. The electron is represented as a particle of mass m and charge $-e$ rotating around a proton of charge $+e$, as shown in Fig. 2.1(a).

The rotating charge forms a current

$$j = \frac{e}{t} = e\frac{v}{2\pi r}$$

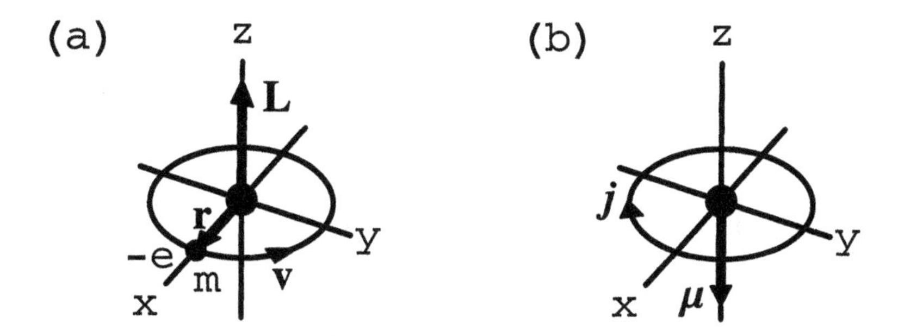

Figure 2.1. An electron rotating around a proton in the Bohr model.

and a *magnetic (dipole) moment*

$$\boldsymbol{\mu}_L = -jr^2\pi\boldsymbol{k} = -\frac{1}{2}e\mathbf{r}\times\mathbf{v} = -\frac{e}{2m}\mathbf{L}, \tag{2.1}$$

where $\boldsymbol{k}$ is the unit vector along the z-axis, $\mathbf{L}$ is the angular momentum of the electron, and m is the electron mass. Therefore, the electron carries a magnetic moment proportional to its angular momentum. Since this outcome is valid in general quantum mechanics, it is valid for the angular momentum operator and therefore not only for the orbital motion of the electron but also for its intrinsic spin angular momentum. The values of the angular momentum are determined by the *angular momentum quantum number*, l, and the values of its z-component by the *magnetic quantum number*, m_l:

$$L = \hbar\sqrt{l(l+1)}, \quad l = 0,\ldots,n-1, \quad L_z = m_l\hbar, \quad m_l = -l,\ldots,l, \tag{2.2}$$

where n is the main quantum number. Introducing $\mu_B = e\hbar/2m$, called the *Bohr magneton*, we can write Eq. (2.1) as

$$\boldsymbol{\mu}_L = -\frac{\mu_B}{\hbar}\mathbf{L}. \tag{2.3}$$

From Eq. (2.2) we also get

$$\mu_{lz} = -\mu_B m_l. \tag{2.4}$$

Similarly, for the electronic spin we have

$$S = \hbar\sqrt{s(s+1)}, \qquad S_z = m_s\hbar, \quad m_s = \pm\frac{1}{2}. \tag{2.5}$$

We do not have a good semiclassical picture for spin (an electron spinning around itself does not work), and therefore we can only conclude that

$$\boldsymbol{\mu}_S = -\frac{g_s e}{2m}\mathbf{S} = -\frac{g_s\mu_B}{\hbar}\mathbf{S}, \qquad \mu_{sz} = -g_s\mu_B m_s, \tag{2.6}$$

where g_s is called the *gyromagnetic factor* (see p. 46) We need another theory—quantum electrodynamics, QED—to determine it: $g_s = 2.002319304386$. We often take it to be 2.

The main idea behind manipulating qubits is to put them in an external field we can control and to get a response we can detect. For our atom, the field would be a uniform magnetic field $\mathbf{B}$ and the response would be spectral lines. In the magnetic field $\mathbf{B}$ oriented along the z-axis, the electron can have the following potential energies:

$$U_L = -\boldsymbol{\mu}_L \cdot \mathbf{B}, \qquad U_S = -g_s\boldsymbol{\mu}_S \cdot \mathbf{B}, \tag{2.7}$$

and from Eq. (2.4) we see that the energies of the electron are changed by the amounts

$$\Delta E_l = m_l\mu_B B, \qquad \Delta E_s = m_l g_s \mu_B B \tag{2.8}$$

from the value they had in the absence of the magnetic field ($B = 0$). This change can be detected by observing the splitting of the spectral lines. For instance, when an electron in an excited state 2p ($n = 2,\ l = 1$) within a hydrogen atom deexcites to the ground state 1s ($n = 1,\ l = 0$) via the emission of a photon, the photon will have only one spectral line, i.e., only one frequency. If we put the atom in a magnetic field, the so-called *normal Zeeman splitting* occurs following Eq. (2.8), as shown in Fig. 2.2.

We can see that there is always an odd number of levels and spectral lines, and therefore we cannot have a one-to-one correspondence between Zeeman levels and states of two-level systems that we need for quantum computation. We could look for just pairs of states as will do in Sec. 3.1.6 (p. 146), but this is not what we would call an off-the-shelf solution. If we try spins we can get an even level splitting, but things get complicated then. To see this let us again consider the semiclassical Bohr picture.

In a hydrogen atom, the electron circles around the proton; but in a system fixed to an electron, the proton circles around the electron and generates the following inner magnetic field at the position of the electron:

$$B_{\text{in}} = \mu_0\frac{I}{2r} = \mu_0\frac{e}{t}\frac{1}{2r} = \mu_0\frac{e}{2r}\frac{v}{2r\pi} = \frac{\mu_0 eL}{4\pi m r^3}. \tag{2.9}$$

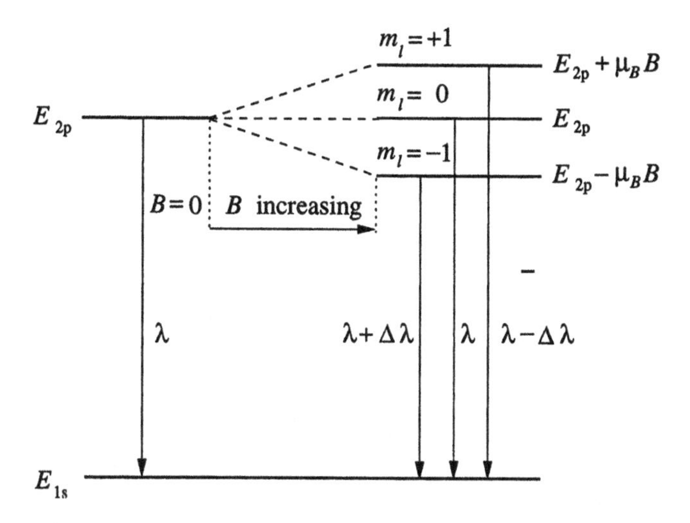

Figure 2.2. The normal Zeeman splitting.

The direction of this inner magnetic field generated by the orbital motion of the electron must be the direction of the angular momentum, **L**. Eqs. (2.9) and (2.6) give the following potential energy for the electron's spin:

$$U_J = -\boldsymbol{\mu}_S \cdot \mathbf{B}_{\text{in}} = \frac{\mu_0 e^2}{4\pi r^3 m^2} \mathbf{S} \cdot \mathbf{L} = \frac{e^2}{4\pi\epsilon_0 r^3 m^2 c^2} \mathbf{S} \cdot \mathbf{L}, \tag{2.10}$$

which is called the *spin–orbit interaction.* Comparing Eqs. (2.7) and (2.10), we interpret the spin–orbit interaction as an internal Zeeman effect. The spectral lines of the corresponding spontaneous emission are split into an *even* number of lines, as shown on the left-hand side of Fig. 2.3 for sodium.

Thus, what characterizes the electrons within an atom is the *total* angular momentum **J**:

$$\begin{aligned} \mathbf{J} = \mathbf{L} + \mathbf{S}, \qquad J = \hbar\sqrt{j(j+1)} \qquad j = |l \pm s|, \\ J_z = m_j \hbar, \qquad m_j = -j, -j+1, \ldots, j. \end{aligned} \tag{2.11}$$

The splitting of spectral lines under the influence of the external magnetic field is called the *anomalous* Zeeman effect and is shown in Fig. 2.3.

The magnetic moment and its z components are

$$\boldsymbol{\mu}_J = -\frac{e}{2m}\mathbf{J}, \qquad \mu_{jz} = -g\mu_B m_j, \tag{2.12}$$

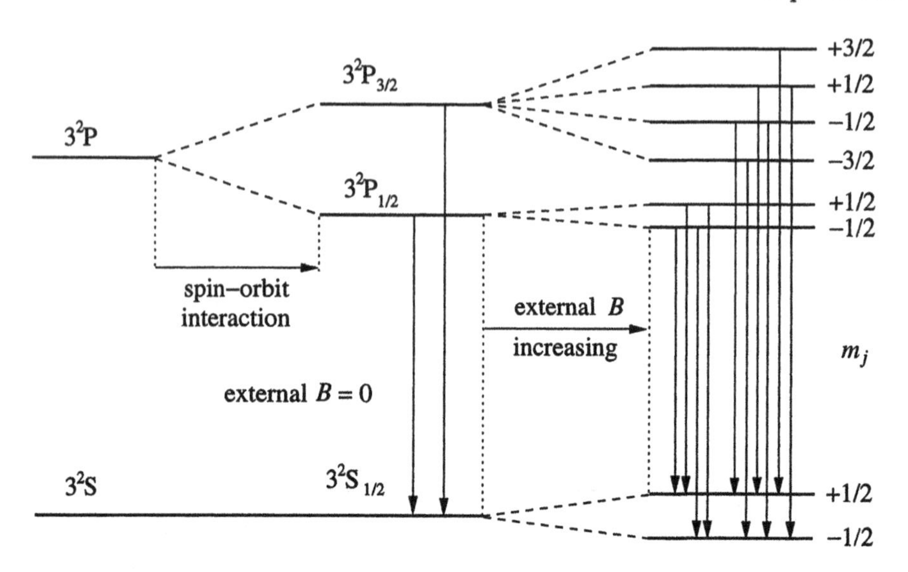

Figure 2.3. The spin–orbit interaction of sodium splits 3^2P into $3^2P_{3/2}$ and $3^2P_{1/2}$, and the external magnetic field splits these and the ground state 3^2S further into the lines shown on the right (ΔE shifts). The selection rules $\Delta j = 0$ ($j \neq 0$), ± 1, $\Delta m_j = 0, \pm 1$, give the spectral lines—an anomalous Zeeman effect.

where

$$g = 1 + \frac{j(j+1) + s(s+1) - l(l+1)}{2j(j+1)} \tag{2.13}$$

is called the *Landé factor* [Greiner, 1989]. The potential energy of the electron in the field is

$$U_J = -\boldsymbol{\mu}_J \cdot \mathbf{B}. \tag{2.14}$$

Due to this magnetic moment, by putting atoms into the magnetic field $\mathbf{B}$, the following energy shift occurs between $B = |\mathbf{B}| = 0, j$ level, and $B \neq 0, j, m_j$ level:

$$\Delta E_j = g m_j \mu_B B. \tag{2.15}$$

However, in deriving this equation [as well as Eq. (2.8)] we assumed that $\mathbf{J}$ ($\mathbf{L}$) is aligned with $\mathbf{B}$. For Eqs. (2.1), (2.12), (2.7), and (2.14) show that when $\mathbf{J}$ ($\mathbf{L}$) is perpendicular to $\mathbf{B}$ we have $U = \Delta E = 0$. $\mathbf{J}$ can be oriented in any direction when we switch on $\mathbf{B}$, and then $\mathbf{B}$ acts on the electron with torque

$$\boldsymbol{\tau}_J = \boldsymbol{\mu}_J \times \mathbf{B} = -g\mu_B \mathbf{J} \times \mathbf{B}. \tag{2.16}$$

If we neglect the spin–orbit interaction, we then have

$$\boldsymbol{\tau}_L = -\mu_B \mathbf{L} \times \mathbf{B} \qquad \text{and} \qquad \boldsymbol{\tau}_S = -g_S \mu_B \mathbf{S} \times \mathbf{B}. \tag{2.17}$$

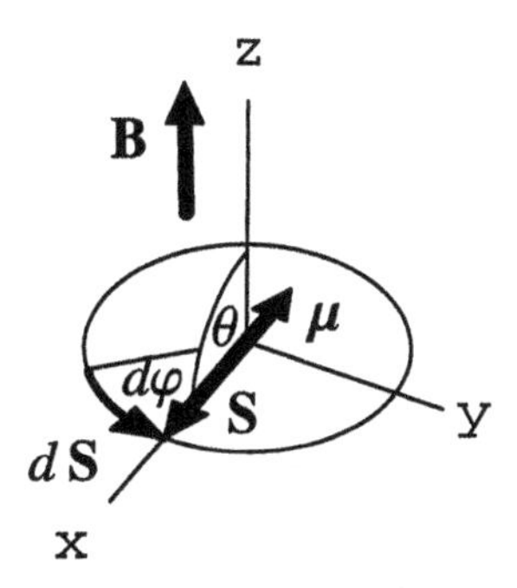

Figure 2.4. Electron with spin precessing in a magnetic field **B**. $\theta = \pi/2$ is chosen although θ can have any value between 0 and π.

The torques will cause changes in the directions of **L** and **S**:

$$\boldsymbol{\tau}_L = -\frac{d\mathbf{L}}{dt}, \qquad \boldsymbol{\tau}_S = -\frac{d\mathbf{S}}{dt} \tag{2.18}$$

that amount to precession about **B**, which frequencies, called *Larmor (angular) frequencies*, are (see Fig. 2.4)

$$\omega_L = \frac{d\phi}{dt} = \frac{|d\mathbf{L}/dt|}{L\sin\theta} = \frac{eLB\sin\theta}{L\sin\theta} = \frac{\mu_B}{\hbar}B, \qquad \omega_S = \frac{g_S\mu_B}{\hbar}B. \tag{2.19}$$

Note that $\omega_S \approx 2\omega_L$. The term *Larmor frequency* is also often used for $\nu = \frac{\omega}{2\pi}$.

Using the Planck formula $E = h\nu = \hbar\omega$, we see that the obtained frequencies give energies that are exactly the energy shifts (2.8) and therefore correspond to high frequencies (electron Larmor frequency, ν for 1 Tesla magnetic field is 28.025 GHz). This means that the detection of such energies for two-level systems, i.e., for spin 1/2 qubits, would be difficult—regardless of whether we attempt to carry out direct detection of the precession energies or indirect detection through the Zeeman splitting of the spontaneous emissions. Then the spontaneous transitions occur within very short time intervals (10^{-8}–10^{-4} sec). In the end, electron spin states are always "screened" by the spin–orbit interaction as well as by the electron interaction with the nucleus, which is several orders of magnitude more massive than electrons.

Nonetheless, we have enough elements to design a system that could perform quantum logic operations on a single qubit, be it electron, nucleus, or the whole atom. An external magnetic field is very easy to apply to a system. After a time interval, one can expect all systems to be aligned with the field. In the initial state, a system does not move—does not precess. Next, we apply perpendicular fields to tip systems that would then start precessing. The energy emitted by the precessing systems should be detectable. Which systems would fit into such a scheme the best: electrons, nuclei, ions, atoms? The two most important requirements are that energies corresponding to precession be much lower than with electrons and that the decoherence time be much longer. Eqs. (2.8) yield $E \sim 1/m$, telling us that for nuclei, which are several orders of magnitude more massive than electrons, we can have radio frequency emissions, and this is something than can be handled easily (the proton Larmor frequency, ν, for a 1-Tesla magnetic field is 42.578 MHz). An off-the-shelf candidate for such a scheme is nuclear magnetic resonance, especially since it also offers very long coherence times.

2.3 Liquid-State Nuclear Magnetic Resonance

Nuclear magnetic resonance (NMR) is a term denoting a "Zeeman-like" method for manipulating and measuring nuclear spins. Nuclei have spins characterized by a *nuclear spin quantum number*, I, and therefore

$$S = \hbar\sqrt{I(I+1)}, \qquad S_z = m_I \hbar, \qquad m_I = -I, \ldots, I. \tag{2.20}$$

The associated intrinsic magnetic moment is

$$\boldsymbol{\mu}_I = \frac{g_N \mu_N}{\hbar}\mathbf{S}, \qquad \mu_{jz} = g_N \mu_N m_j, \tag{2.21}$$

where $\mu_N = \frac{e\hbar}{2m_p}$ (m_p being the proton mass) and g_N is the *nuclear g factor*. Nuclear magnetic moments are usually specified by the values of their g factors, which can be positive and negative. For instance, for the proton we have $g_p = 5.5856912 \pm 0.0000022$. The neutron also has a nuclear magnetic moment, and its g factor is $g_n = -3.8260837 \pm 0.0000018$. There is no direct semiclassical interpretation of the nuclear magnetic moment for the neutron, although classical charges that sum up to a net charge of zero can have a magnetic dipole moment. The g factors differ greatly from one nuclide to another. For example, ^{17}O has $g = -0.76$, ^{93}Nb has 2.47, and ^{57}Fe has 0.18. Note that $\boldsymbol{\mu}_N$ and $\mathbf{S}$ are parallel for positive and antiparallel for negative g factors.

When we put a nucleus (coupling to electrons within an atom can be treated as a small perturbation) into a static magnetic field B, we will get energy shifts analogous to those from Eqs. (2.8) and (2.15):

$$\Delta E_I = g_N \mu_N m_I B. \tag{2.22}$$

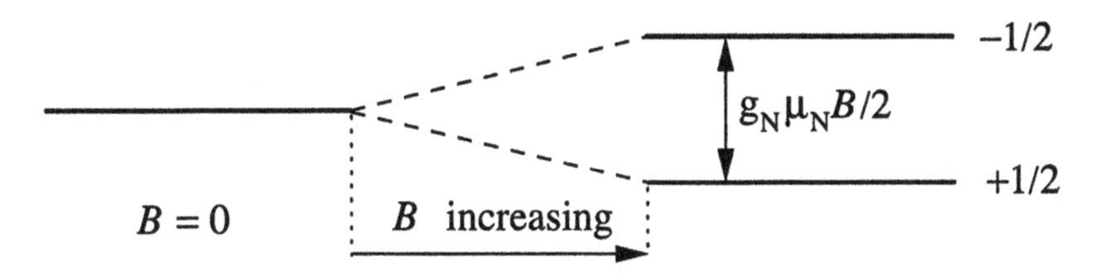

Figure 2.5. The nuclear Zeeman effect for a spin-$\frac{1}{2}$ nucleus in a magnetic field **B** for a positive nuclear g factor (for negative g, the levels are inversed).

As a result, we obtain the *nuclear Zeeman effect* shown in Fig. 2.5.

The nuclear Zeeman effect differs from its atomic version (presented in Sec. 2.2) in two respects. First, we can handle spins directly, since they are not strongly bound to another observable, unlike spin-orbit electron coupling. Second, we can directly measure ΔE_I given by Eq. (2.22), as opposed to the electron case, where we obtain a shift in a spectral line (obtained by a spontaneous transition between states determined by other observables (cf. 2.3)); actually, spontaneous transitions take place between nuclear states as well, but the probability of their occurrence as compared to the stimulated transition is negligible.

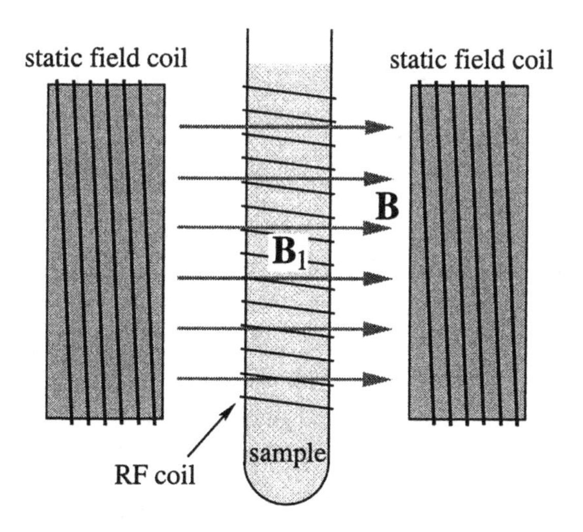

Figure 2.6. Experimental setup for NMR preparation and measurement.

By means of the device shown schematically in Fig. 2.6, we can handle nuclear spins (within a sample containing, for instance, chloroform, alanine, or trichloroethylene) using a weak rotating magnetic field $\mathbf{B}_1$ and a strong static field $\mathbf{B}$ (typically $\mathbf{B}_1 < 0.001\mathbf{B}$). In real experiments, one does not use a rotating field but an oscillating field generated by an alternating *radio-frequency (RF)* current passing through an *RF coil.* An oscillating field is a superposition of two oppositely rotating fields.

To better understand how this process works, let us first assume that the magnetic moment $\boldsymbol{\mu}_I$—corresponding to qubit $|0\rangle$—is oriented along the z-axis. Then we briefly apply $\mathbf{B}_1$ along the x-axis to the magnetic moment and tip it a little off the z-axis. Due to the field $\mathbf{B}$, it starts precessing with frequency

$$\omega_I = \boldsymbol{\mu}_I \cdot \mathbf{B} = \frac{g_N \mu_N}{2\hbar} B, \tag{2.23}$$

where 1/2 stands for $|m_I| = 1/2$. We continue to apply $\mathbf{B}_1$ in pulses at the frequency ω_I. We call these pulses *resonant pulses.* In the frame rotating around the z-axis at the frequency ω_I, the field $\mathbf{B}_1$ looks like a constant vector, for example, along the x-axis as shown in Fig. 2.7.

When $\boldsymbol{\mu}_I$ is tipped to the xy-plane (see Fig. 2.7 (a); we have to keep the *RF generator* on for time $t = \frac{\pi}{2\omega_I}$), the precession in the xy-plane induces an RF signal in a pickup coil (see Fig. 2.7(b); it can be a separate coil as shown here or the same coil that produces $\mathbf{B}_1$, as in Fig. 2.6).

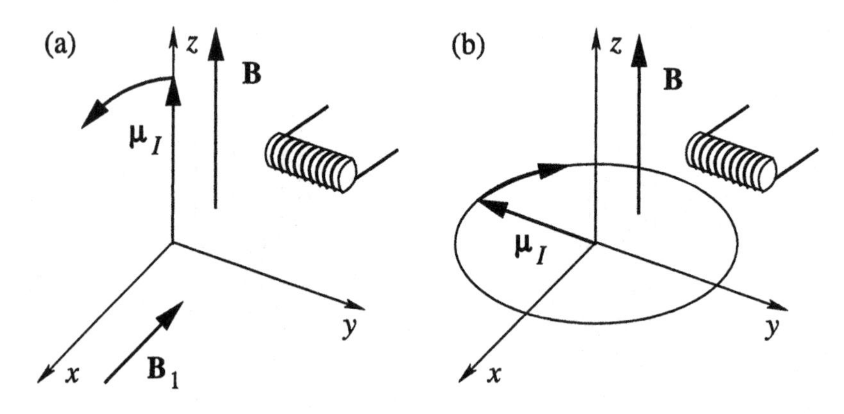

Figure 2.7. In a rotating frame, $\mathbf{B}_1$ is aligned with the x-axis and tips $\boldsymbol{\mu}_I$ into xy-plane. Then the *precession* about the z-axis induces a signal in the coil.

The signal induced by a single nucleus would be too weak and could not be measured. Consequently, we use liquid samples (usually called *lattices*). Each molecule in such a large ensemble of molecules is simultaneously placed in an initial state and subsequently subjected to RF pulses. In the end, these all induce RF signals in a *pickup coil* that sum up to a detectable output.

Different nuclei have different resonant frequencies because they have different g factors, g_N. Hence an NMR experiment can prepare and detect states of different nuclei, i.e., different qubits. These different qubits must be coupled, and for this purpose appropriate molecules are chosen. We have already mentioned some of them above. For instance, trichloroethylene,

Cl, Cl
$^{13}C = {}^{13}C$
H Cl
,

can incorporate three qubits: the spin-$\frac{1}{2}$ proton, H; and two spin-$\frac{1}{2}$ ^{13}C carbons (the usual ^{12}C ethylene carbons have spin 0) [Laflamme et al., 2002].

The qubits correspond to the $-\frac{1}{2}$ and $+\frac{1}{2}$ energy levels shown in Fig. 2.5 and therefore to magnetic moments that point up and down respectively:

$$|\psi_t\rangle = e^{-i\omega t/2}\alpha|0\rangle + e^{i\omega t/2}\beta|1\rangle, \tag{2.24}$$

where ω is the precession frequency given by Eq. (2.23). For the above trichloroethylene at $B = 11.7$ T, the precession frequency for protons (i.e., for H) is about 500 MHz and about 124.5 MHz and 125.5 MHz for the two ^{13}C frequencies.

By means of σ_z (given by Eq. (1.88)) aligned with the field B, we can write Eq. (2.24) as

$$|\psi_t\rangle = e^{i\omega\sigma_z t/2}|\psi_0\rangle. \tag{2.25}$$

The equation can be verified with the help of MatrixExp [cf. Eq. (1.108)]. The observable σ_z measures the spin along the z-axis. Analogously, σ_x and σ_y measure the spin along the x and y axes determined by the orthogonal weak fields $\mathbf{B}_1$ and $\mathbf{B}_2$ produced by RF coils (see Fig. 2.6, p. 95). Hence, $|\psi_t\rangle$ is the function $|\Psi\rangle$ given by Eq. (1.113), which represents a state on the Bloch sphere. The corresponding density matrix is given by Eq. (1.112), which can also be expressed as Eq. (1.92), where $\mathbf{r}$, given by Eq. (1.111), points to the states on the Bloch sphere.

Thus, one-qubit gates are rotations within the Bloch sphere implemented by RF pulses. This is described by the Hamiltonian

$$H = \frac{1}{2}(\omega_x\sigma_x + \omega_y\sigma_y + \omega_z\sigma_z), \tag{2.26}$$

and the corresponding state is

$$|\psi_t\rangle = e^{-iHt}|\psi_0\rangle. \tag{2.27}$$

For instance, by applying an RF pulse along the x-axis with duration $t = \pi/(2\omega_x)$, we get $\sqrt{\text{NOT}}$ up to a phase (cf. Eq. (1.56))

$$e^{-i\pi\sigma_x/4} = \text{MatrixExp}[-\frac{i\pi\sigma_x}{4}] = \frac{1}{\sqrt{2}}\begin{bmatrix} 1 & i \\ i & 1 \end{bmatrix}. \tag{2.28}$$

Similarly, we can get the Hadamard gate (Eq. (1.59)) up to a phase $\pi/2$:

$$e^{-i\pi\sigma_x/2}e^{-i\pi\sigma_y/4} = \frac{i}{\sqrt{2}}\begin{bmatrix} 1 & 1 \\ 1 & -1 \end{bmatrix}. \tag{2.29}$$

With the help of single qubit gates and CNOT gates, we can implement almost any gate and algorithm (see Sec. 1.19, p. 51; and Theorem 1.3, p. 53). Therefore, what is left to be considered are two gates manipulated so as to give a CNOT gate. To be able to form a CNOT gate, two gates must be coupled—the so-called *J-coupling* in NMR. J-coupling is essentially a spin–spin magnetic interaction, although distorted by its electron mediation and the molecule motion. It can be approximated by the Hamiltonian

$$H_J = \frac{\pi J}{2}\sigma_{1z}\sigma_{2z}, \tag{2.30}$$

where σ_i, $i = 1, 2$, are the Pauli operators of the nuclei and J is the *coupling constant.* J is about 100 Hz between the two ^{13}C atoms and about 100 Hz between H and the adjoining ^{13}C atoms in trichloroethylene.

J-coupling causes an increase in the precession frequency of one of the spins when the other is oriented along the $+z$-axis ($|0\rangle$ state), and a decrease when the latter spin is oriented along the $-z$-axis ($|1\rangle$ state). As a result, the former spin rotates anticlockwise in the rotating xy-plane when the latter is in the state $|0\rangle$, and clockwise when it is in the state $|1\rangle$, as shown in Fig. 2.8 (a3) and (b3).

Actually, virtually all basic quantum gates and algorithms have been implemented in NMR devices, including Shor's algorithm for factoring numbers (the number 15 was factored by means of a 7-qubit molecule [Vandersypen et al., 2001]). We can say that NMR devices, on which dozens of experiments have been carried out so far, are the first fully operating quantum computers. They are also well ahead of any other existing model and therefore deserve special attention. Yet these devices are apparently not prototypes for realistic would-be quantum computers, since we can hardly scale them up to a large enough number of qubits to produce machines of any useful size.[1] The first problem is, of course,

[1] It has been estimated that we would need a 1000-qubit quantum computer to outperform a classical computer [Preskill, 1998].

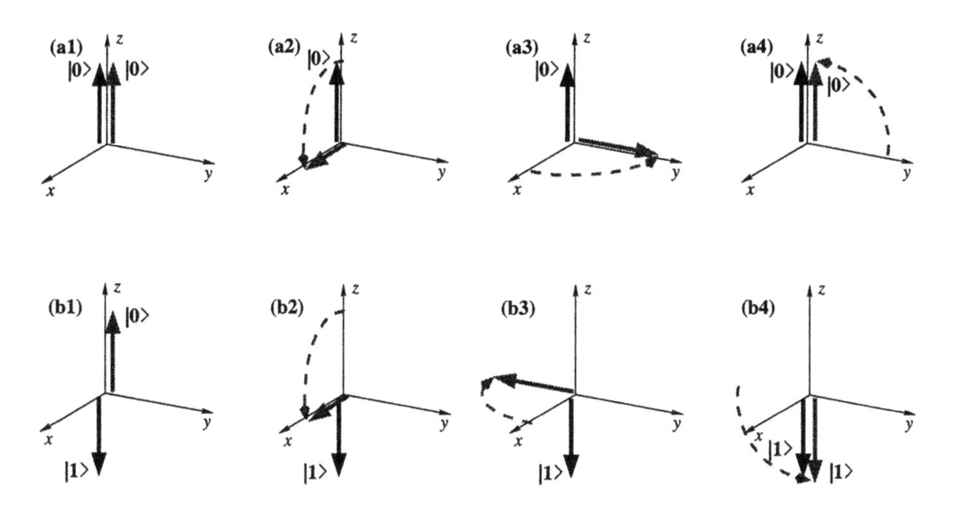

Figure 2.8. CNOT NMR gate. The a-row shows CNOT$|0\rangle|0\rangle$=$|0\rangle|0\rangle$ and the b-row shows CNOT$|1\rangle|0\rangle$=$|1\rangle|1\rangle$. The black arrow is a control qubit, and the gray arrow is a target qubit. We arrive at (a2) and (b2) by y-rotation (RF pulses). (a3) and (b3) are the consequences of the J-coupling. After applying x-rotations, we arrive at (a4) and (b4). The figure is according to Fig. 11 of [Laflamme et al., 2002].

the size of the molecules. Another problem is the operating temperature and the strength of the magnetic field, because at higher temperatures, the more qubits we have, the exponentially bigger the error. For low temperatures, Eq. (2.22) and basic thermodynamics yield $\mu_I B \sim kT$. But the strongest available NMR field today (less than 30 Tesla) corresponds to $T \approx 0.04$ K, and at this temperature the sample cannot be in the liquid state. There are many other problems with scaling, such as increasing the frequencies and decoherence. Nuclear spin quantum computer models and experiments must therefore be scaled in a different way, for example, by using a solid state, as we describe in the next section. There might also be a fundamental problem with the present implementations of NMR computing: it is not obvious whether entanglement, which can be obtained in principle, is also obtainable in realistic implementations where apparently all room-temperature thermal states should be separable [Braunstein et al., 1999].

2.4 Silicon-Based Nuclear Spins

The scalability problem that NMR quantum computers have with the size of molecules stems from the fact that we address nuclei by their chemical identity. We do so with the help of RF coils tuned to the resonant frequencies characteristic of the addressed nuclei. This is apparently simply not feasible for molecules with more than 20 qubits. Hence, it would be very convenient if we could find a way to address nuclei by their addresses as in classical computers. It would be even more convenient if we could take advantage of the existing research on the materials used by the standard computer industry, which currently approaches the quantum realm from the "other side." And this is exactly what recent silicon-based nuclear spin computer models offer.

Nuclear spins in solid state materials interact with electron spins in a way similar to that in which electron angular momenta interact with electron spins through the spin–orbit interaction (Eq. (2.10) and Fig. 2.3), i.e., the nuclear magnetic moment interacts with the electron magnetic moment. This interaction—called the *hyperfine interaction*—contributes to the Hamiltonian of the whole system containing nuclei and electrons with the term

$$H_h = A\mathbf{I}\cdot\mathbf{S}, \tag{2.31}$$

where $\mathbf{S}$ and $\mathbf{I}$ are the electron and nuclear spins, respectively (cf. Sec. 2.3, p. 94).

With $S = \frac{1}{2}$ and $I = \frac{1}{2}$, the base states for the nucleus and the electron taking part in the hyperfine interaction are $|\uparrow\rangle_e|\uparrow\rangle_N$, $|\uparrow\rangle_e|\downarrow\rangle_N$, $|\downarrow\rangle_e|\uparrow\rangle_N$, and $|\downarrow\rangle_e|\downarrow\rangle_N$. Their arbitrary state is therefore

$$|\Psi\rangle = C_1|\uparrow\rangle_e|\uparrow\rangle_N + C_2|\uparrow\rangle_e|\downarrow\rangle_N + C_3|\downarrow\rangle_e|\uparrow\rangle_N + C_4|\downarrow\rangle_e|\downarrow\rangle_N. \tag{2.32}$$

The action of the spin operators on the base states is the standard Pauli matrix action on them. For example, $S_y|\uparrow\rangle_e|\downarrow\rangle_N = \sigma_y^e|\uparrow\rangle_e|\downarrow\rangle_N = i|\downarrow\rangle_e|\downarrow\rangle_N$, $I_x|\uparrow\rangle_e|\downarrow\rangle_N = \sigma_x^N|\uparrow\rangle_e|\downarrow\rangle_N = |\uparrow\rangle_e|\uparrow\rangle_N$, and $S_xI_z|\uparrow\rangle_e|\downarrow\rangle_N = \sigma_x^e\sigma_z^N|\downarrow\rangle_e|\uparrow\rangle_N = -|\downarrow\rangle_e|\downarrow\rangle_N$. The total Hamiltonian is a constant plus the interaction part given by Eq. (2.31):

$$H = E_0\mathbb{1} + A\,\boldsymbol{\sigma}^e\cdot\boldsymbol{\sigma}^N. \tag{2.33}$$

Using Eq. (2.6) and (2.21), we can also understand this interaction as the interaction of two magnetic dipoles whose energy depends on $\boldsymbol{\mu}_S\cdot\boldsymbol{\mu}_I$. The difference between the classical and quantum dipole interactions is that the classical interaction depends on the distance between the dipoles, while the quantum interaction does not—the Hamiltonian (2.33) gives only the average interaction energy [Feynman et al., 1965, 12-5]. And this is the key point for our application because in the solid state, in particular in semiconductors, the electron wave function is spread through the crystal lattice surrounding a nucleus or ensemble of nuclei.

Resolving $|\Psi\rangle$ into base states, considering its time evolution, and taking into account that $C_i = C_i(t) = \langle\Psi_i(t)|\Psi(t)\rangle = \langle\Psi_i|\Psi\rangle$, $i = 1,\dots,4$, where $|\Psi_i\rangle$ is one of the four base states from Eq. (2.32), we get [Feynman et al., 1965, 8-9]

$$i\hbar\frac{dC_i}{dt} = \sum_{j=1}^{4} H_{ij}C_j, \qquad i = 1, \ldots, 4, \tag{2.34}$$

where $H_{ij} = \langle i|H|j\rangle$. We obtain these elements by applying H to all base states. For instance,

$$H|\uparrow\uparrow\rangle = A\,\boldsymbol{\sigma}^e\cdot\boldsymbol{\sigma}^N = A(\sigma_x^e\sigma_x^N + \sigma_y^e\sigma_y^N + \sigma_z^e\sigma_z^N)|\uparrow\uparrow\rangle = A\,|\uparrow\uparrow\rangle, \ldots,$$

where we adopt the notation $|\uparrow\uparrow\rangle$ for $|\uparrow\rangle_e|\uparrow\rangle_N$, etc. This yields

$$H_{11} = \langle\uparrow\uparrow|H|\uparrow\uparrow\rangle = A, \quad H_{12} = \langle\uparrow\downarrow|H|\uparrow\uparrow\rangle = 0, \ldots, \tag{2.35}$$

and taking

$$C_i = a_i e^{-i\omega t} \tag{2.36}$$

we get the following system:

$$\begin{aligned} Ea_1 &= Aa_1, & Ea_2 &= -Aa_2 + 2Aa_3, \\ Ea_3 &= Aa_2 - 2Aa_3, & Ea_4 &= Aa_4. \end{aligned} \tag{2.37}$$

From the first and the fourth equations, we get the following energies and states:

$$E_I = A, \quad |I\rangle = |\uparrow\uparrow\rangle, \quad E_{II} = A, \quad |II\rangle = |\downarrow\downarrow\rangle, \tag{2.38}$$

and from the second and the third we get

$$\begin{aligned} E_{III} &= A, & |III\rangle &= \frac{1}{\sqrt{2}}(|\uparrow\downarrow\rangle + |\downarrow\uparrow\rangle), \\ E_{IV} &= -3A, & |IV\rangle &= \frac{1}{\sqrt{2}}(|\uparrow\downarrow\rangle - |\downarrow\uparrow\rangle). \end{aligned} \tag{2.39}$$

Hence, the states $|I\rangle$, $|II\rangle$, and $|III\rangle$ are degenerate, and a transition from them to the state $|IV\rangle$ causes an emission of a microwave quantum, $\Delta E = 4A = \hbar\omega$. Of course, the atom would also absorb a quantum of the frequency ω.

When we put the atom into an external magnetic field $\mathbf{B}$, the upper line splits. Let us look at the details, following [Feynman et al., 1965, 12-9]. Feynman presented the calculations only for hydrogen, but they approximate any atom with nuclear spin $\frac{1}{2}$ and a single electron in the outer shell [Kane, 2000]. The Hamiltonian is

$$H = A\,\boldsymbol{\sigma}^e\cdot\boldsymbol{\sigma}^N + g_S\mu_B\boldsymbol{\sigma}^e\cdot\mathbf{B} - g_N\mu_N\boldsymbol{\sigma}^e\cdot\mathbf{B}, \tag{2.40}$$

where g_N, μ_N, and $g_S\mu_B$ are given by Eqs. (2.6) and (2.21). Proceeding in the same way as above and obtaining equations analogous to Eqs. (2.35)–(2.37), we get the following energies:

$$\begin{aligned} E_I &= A + (g_S\mu_B - g_N\mu_N)B, \\ E_{II} &= A - (g_S\mu_B - g_N\mu_N)B, \\ E_{III} &= A\,(-1 + 2\sqrt{1 + (g_S\mu_B + g_N\mu_N)^2 B^2/4A^2}\;), \\ E_{IV} &= -A\,(1 + 2\sqrt{1 + (g_S\mu_B + g_N\mu_N)^2 B^2/4A^2}\;). \end{aligned} \tag{2.41}$$

Since $g_S\mu_B$ (for the electron) is about thousand times larger than $g_N\mu_N B$ (for the nucleus), we get the plot shown in Fig. 2.9.

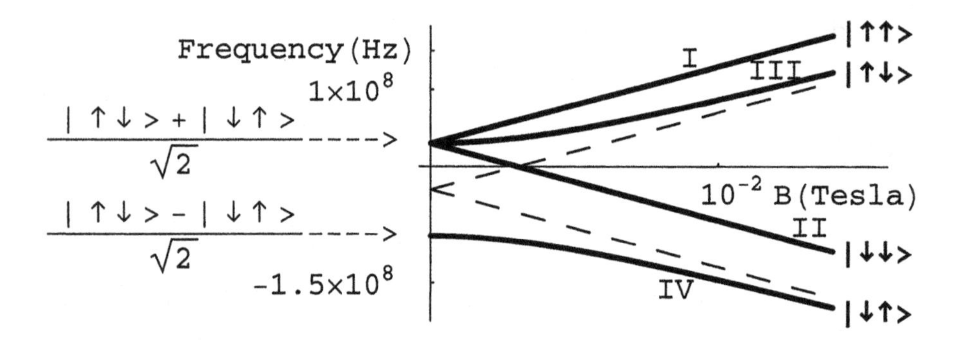

Figure 2.9. The energy levels (Eqs. (2.41)) and states of an atom with a nucleus of spin $\frac{1}{2}$ and a single electron in the highest shell with a negligible spin-orbit interaction. The dashed lines are $-A+(g_S\mu_B+g_N\mu_N)B$ and $-A-(g_S\mu_B+g_N\mu_N)B$. All emission and absorption transitions are allowed.

From Eq. (2.41) we can see that for large B, the dashed linear functions given in Fig. 2.9 approximate E_{III} and E_{IV}. Thus, we obtain four linear functions representing four energies and four different states $|I\rangle = |\uparrow\uparrow\rangle$, $|II\rangle = |\downarrow\downarrow\rangle$, $|III\rangle = C_{III2}|\uparrow\downarrow\rangle + C_{III3}|\downarrow\uparrow\rangle$, and $|IV\rangle = C_{IV2}|\downarrow\uparrow\rangle - C_{IV3}|\downarrow\uparrow\rangle$, where C_j are (for $B > 0$ and Hamiltonian (2.40)) determined by analogy with Eq. (2.34). It is straightforward to show [Feynman et al., 1965] that for small magnetic fields, $C_{III2}{=}C_{III3}{=}C_{IV2}{=}C_{IV3}{=}1/\sqrt{2}$ and for large ones, $C_{III2} = 1$, $C_{III3} \approx 0$, $C_{IV2} \approx 0$, and $C_{IV3} = 1$.

These properties of atoms with nuclear spin $\frac{1}{2}$, which have no orbital degree of freedom and have a single electron in the highest shell with a negligible spin–orbit interaction, promise comparatively easy handling within a solid state environment. The natural candidate for the atom is phosphorus, which appears in only one isotope ^{31}P in nature, and therefore is 100% pure. It has one outer electron, and $I = \frac{1}{2}$. The natural solid state candidate is the group IV of Si semiconductors with $I = 0$. Current intensive efforts to shrink silicon processors may yield many techniques that might prove helpful for implementation of qubits into an Si semiconductor environment. Recently Bruce Kane proposed such an implementation [Kane, 1998; Kane, 2000; Skinner et al., 2003]. Let us consider the details.

The Si substrate is an insulator at temperatures under 1 K, i.e., there are no free electrons. Phosphorus atoms can be introduced as donors in controllable positions in the substrate. As we have pointed out above, the energy levels of the hyperfine interaction between the electron and nuclear spins in ^{31}P are well defined, and the transitions between the levels shown in Fig. 2.9 are induced by a globally applied radio frequency magnetic field B_{RF}. Bringing the systems into resonance with B_{RF} is done by *A-gates* as shown in Fig. 2.10.

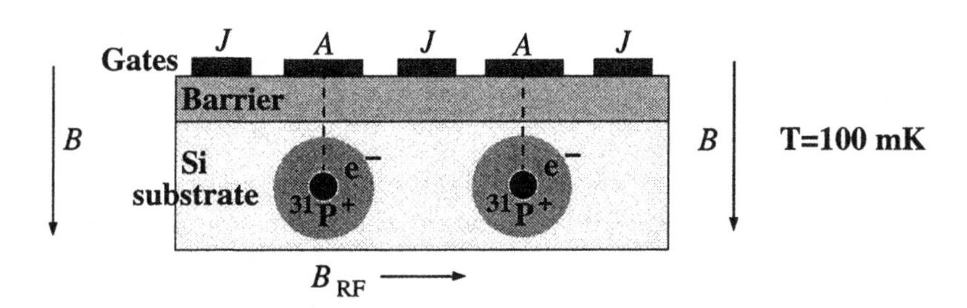

Figure 2.10. Phosphorus donors in an Si substrate. *A*-gates control the nuclear-resonance frequency, and *J*-gates control the electron-mediated coupling between the nuclear spins. The figure is according to Fig. 1 of [Kane, 1998].

A-gates are small electrodes with the help of which one can address qubits in the following way. In contrast with NMR, where each qubit has a different chemical identity and therefore a different Larmor resonance frequency, the process of addressing the nuclei by means of the external *RF field* can be carried out by only a single frequency, because all qubits have the same chemical identity (^{31}P) and all see the same chemical environment (silicon). Hence, we cannot set the external RF field $B_{\rm RF}$ to this frequency because it would tip the spins of all qubits indiscriminately. The solution is to apply a slightly detuned external $B_{\rm RF}$ and to tune in a chosen qubit instead of the field itself. We do so by means of the *A*-gate voltage over the chosen nucleus. It draws the electron off the nucleus, as shown in Fig. 2.11, thereby changing the resonant frequency of the nucleus so as to coincide with that of the external fields $B_{\rm RF}$.

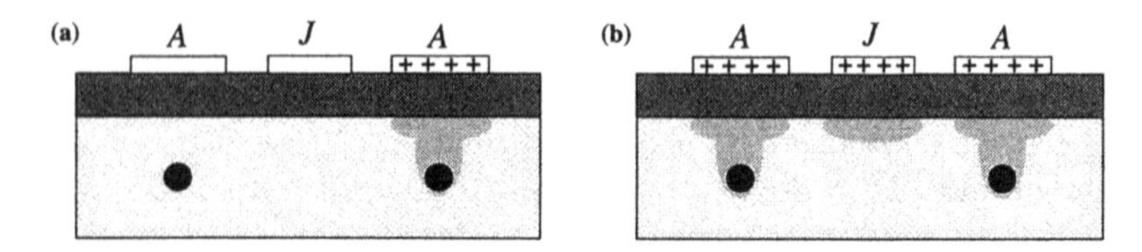

Figure 2.11. (a) One qubit operation: A-gate bias brings the nuclear spin into resonance with external RF field; (b) Two-qubit operation: a positive voltage bias applied to the J-gate lowers the potential barrier between donor sites and turns on exchange electron-mediated coupling between the nuclei.

Two nuclear spins from two adjacent ^{31}P atoms can interact with the same electron and therefore be coupled through an electron mediation. Such an electron-mediated interaction can therefore be controlled by voltages applied to metallic gates above the Si substrate. In Fig. 2.10, this is done by *J-gates.* At sufficiently low temperatures (about 1 K), the electron spin relaxation time is over 15 minutes and the nuclear spin relaxation is over 10 hours, so that there is no problem with decoherence. With even lower temperatures, arbitrarily long relaxation times can be achieved.

For a computation, we can use nuclear spins as memory and electron spins as mediators of interaction between the nuclear spins. There are several reasons for taking this approach. The relaxation time of nuclear spins is much longer, they have no orbital degrees of freedom, and they rotate much more slowly in the external magnetic field B than electrons.

The Hamiltonian of two coupled nucleus–electron systems is

$$H = H_B + A_1\, \boldsymbol{\sigma}^{1e} \cdot \boldsymbol{\sigma}^{1N} + A_2\, \boldsymbol{\sigma}^{2e} \cdot \boldsymbol{\sigma}^{2N} + J\, \boldsymbol{\sigma}^{1e} \cdot \boldsymbol{\sigma}^{2e}, \tag{2.42}$$

where H_B is the part containing the magnetic field interaction terms as in Eq. (2.40); A_1, A_2 are the hyperfine interaction energies; and J is the exchange energy, depending on the degree to which the electron wave function overlaps. The second and third terms in the Hamiltonian H above are called the H_A part [Kane, 1998; Kane, 2000]. The H_B part is also called the *Zeeman magnetic* part [Levy, 2002]. Again, as with Eq. (2.41), we can use the calculations for the hydrogen [Kane, 2000] to get

$$J(r) \sim E_b \left(\frac{r}{a_B}\right)^{\frac{5}{2}} \exp\left(-\frac{2r}{a_B}\right), \tag{2.43}$$

where r is the distance between the donors and a_B is the semiconductor Bohr radius. J can be varied by electrostatic potentials imposed by J-gates positioned between donors, as shown in Figs. 2.10 and 2.11.

We should emphasize that the H_B part in Eq. (2.42) contains terms that involve a strong external field B (which should be called B_z, but it is usual in the literature to simply denote it B),

$$-g_N\mu_N B_z(\sigma_{1N}^z + \sigma_{2N}^z) + g_S\mu_B B_z(\sigma_{1e}^z + \sigma_{2e}^z), \qquad (2.44)$$

and an oscillating field B_{RF} (often also denoted B_{ac}), which is positioned in the xy-plane)

$$\begin{aligned} &B_{\mathrm{RF0}}(t)\cos\omega t[-g_N\mu_N(\sigma_{1N}^x + \sigma_{2N}^x) + g_S\mu_B(\sigma_{1e}^x + \sigma_{2e}^x)] \\ &+B_{\mathrm{RF0}}(t)\sin\omega t[-g_N\mu_N(\sigma_{1N}^y + \sigma_{2N}^y) + g_S\mu_B(\sigma_{1e}^y + \sigma_{2e}^y)]. \end{aligned} \qquad (2.45)$$

A negative voltage bias applied to the J-gate decouples the adjacent spins, and a positive voltage couples them. A positive voltage bias applied to the A-gate draws the electron away from the nucleus, thus reducing the interaction between the electron and nuclear spins and therefore the energy difference between $|\uparrow\rangle_N$ and $|\downarrow\rangle_N$. This outcome allows us to make an arbitrary transition between these states by achieving a resonance with a global RF oscillating magnetic field B_{RF}. Actually, by applying an appropriate voltage fluctuation to A-gates we can rotate—similarly to the NMR rotations presented in Sec. 2.3—the qubits (nuclear spins) and thus change the levels and states (see Fig. 2.9, p. 102). This suffices for handling individual qubits as well as controlled operations. For example, the CNOT of the electron spin, conditioned on the state of the nuclear spin [Kane, 2000], can be achieved by exciting the transition between the lower two states (see Fig. 2.9, p. 102) with the radio frequency magnetic field B_{RF}, provided that the field B is strong enough: $(|\downarrow\rangle_e|\uparrow\rangle_N) \leftrightarrow |\uparrow\rangle_e|\uparrow\rangle_N$. However, if we want to scale up the number of qubits then we have to choose a different approach. But let us first say a few words about reading off the results of a computation.

To read out the states of qubits in which a finished calculation leaves them, we use—unlike NMR computers—charge measurements. Measurements of the magnetic fields of electrons and nuclei would also be a possibility, but these are much slower than the charge measurements. Single charge measurements can be carried out by a *single electron transistor*, SET, in microseconds as we mentioned in Sec. 1.5 (p. 12). Here is yet another example of how recent advances in the shrinking of classical computers can help us in building a solid state quantum computer.

Measurements of nuclear spins are made by *charge measurements*[2] of electrons for the exchange energy $J > \mu_B B/2$. (Computations are made

[2]The SETs are positioned above atoms and the charge motion within Si substrate (see Fig. 2.13) change the potential of a SET part called *island*. This changes SET's conductance and enables measurements.

for $J < \mu_B B/2$ because for lower J, the electrons are fully polarized, while for $J > \mu_B B/2$, the energy levels are split (cf. Fig. 2.9, p. 102). Then the state $(|\uparrow\rangle_{1e}|\downarrow\rangle_{2e} - |\downarrow\rangle_{1e}|\uparrow\rangle_{2e})/\sqrt{2}$ has the lowest energy and $|\uparrow\rangle_{1e}|\uparrow\rangle_{2e}$ has the next to lowest. These two states are coupled to the nuclear spin states, which then determine whether the electrons will be in the state $|\uparrow\rangle_{1e}|\uparrow\rangle_{2e}$ or in $(|\uparrow\rangle_{1e}|\downarrow\rangle_{2e} - |\downarrow\rangle_{1e}|\uparrow\rangle_{2e})/\sqrt{2}$. Since the charge measurement can differentiate between these two electron states, we can infer the values of the nuclear spin states. Thus we can recover both electron spin and nuclear spin. At the same time, this outcome defines the qubits we can work with. The qubits are given by [Kane, 2000]

$$|0\rangle \equiv |\uparrow\rangle_e|\downarrow\rangle_N + |\downarrow\rangle_e|\uparrow\rangle_N, \qquad |1\rangle \equiv |\uparrow\rangle_e|\downarrow\rangle_N - |\downarrow\rangle_e|\uparrow\rangle_N. \tag{2.46}$$

Two-qubit systems and the operations (gates) one can carry out on them are determined by the Hamiltonian given by Eqs. (2.42), (2.44), and (2.44). The calculations of the energy levels for a two-qubit system are straightforward although tedious. They can be done in the same way that the calculations for the one-qubit system, described by Hamiltonian (2.40), were carried out above. Thus, we get equations analogous to Eq. (2.41), and we get Fig. 2.12, which corresponds to Fig. 2.9 (p. 102).

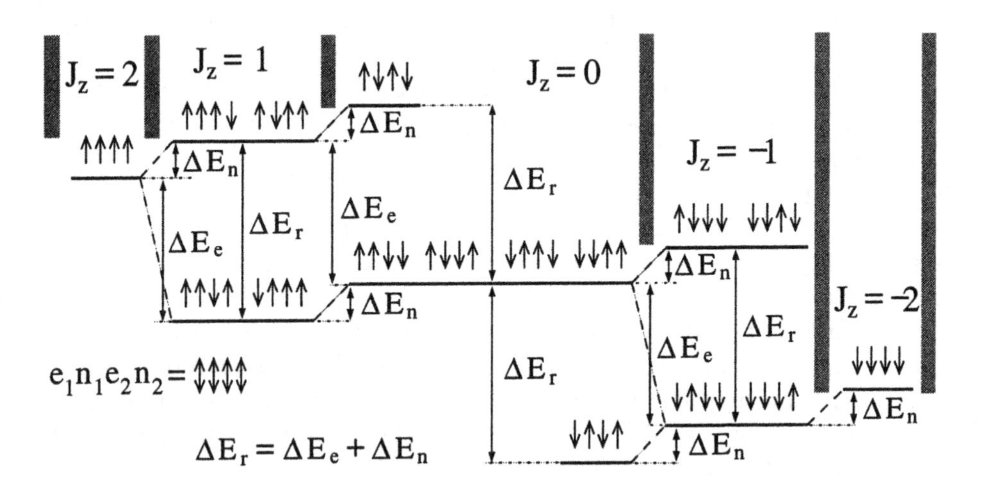

Figure 2.12. Magnetic energy levels of the total spin subspaces of the two-qubit Kane computer. Flipping the electron or nuclear spin changes the energy by ΔE_e or ΔE_N, respectively. Flipping both of them changes the energy by ΔE_r [Skinner et al., 2003].

Now, in order to carry out, for example, the CNOT operation

$$\begin{aligned}\mathrm{CNOT}|1\rangle_1|0\rangle_2 &= \mathrm{CNOT}(|\uparrow\rangle_{1e}|\downarrow\rangle_{1N} - |\downarrow\rangle_{1e}|\uparrow\rangle_{1N}) \\ &\qquad (|\uparrow\rangle_{2e}|\downarrow\rangle_{2N} + |\downarrow\rangle_{2e}|\uparrow\rangle_{2N}) \\ &= (|\uparrow\rangle_{1e}|\downarrow\rangle_{1N} - |\downarrow\rangle_{1e}|\uparrow\rangle_{1N})(|\uparrow\rangle_{2e}|\downarrow\rangle_{2N} - |\downarrow\rangle_{2e}|\uparrow\rangle_{2N}) \\ &= |1\rangle_1|1\rangle_2, \end{aligned} \tag{2.47}$$

we have to find a function that will change the phase of $|\downarrow\rangle_{2e}|\uparrow\rangle_{2N}$ depending on the phase of $|\downarrow\rangle_{1e}|\uparrow\rangle_{1N}$. We do this by allowing electrons to act on the nucleus from the other qubit by means of the hyperfine interaction. Kane designed a hyperfine evolution for this purpose [Skinner et al., 2003].

The idea is to apply *bit trains* of voltage pulses applied to A-gates while S-*gates* shuttle electrons from and to donors. (S-gates correspond to the J-gates in Figs. 2.10 and 2.11.)

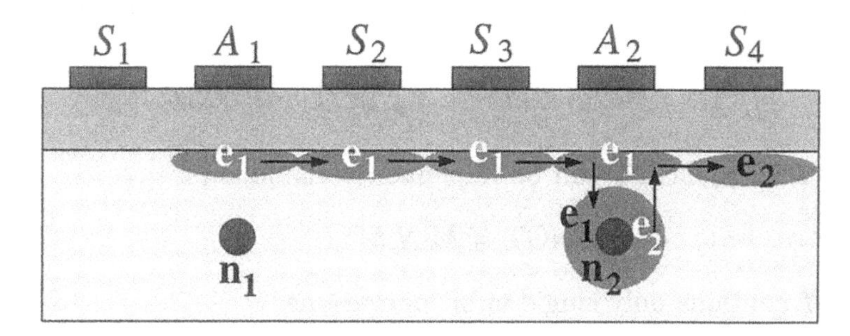

Figure 2.13. Entangling electron-nucleus qubits e_1N_1 and e_2N_2. S-gates displace e_2 and shuttle e_1 from N_1 to N_1. A-gates draw electrons away from nuclei and change their states enough for the global RF field B_{RF} to flip targeted sub-states [Skinner et al., 2003].

First, we apply a positive voltage to A-gate A_2 to draw the electron e_2 off the donor ^{31}P atom, and then, by applying a positive voltage to S-gate S_4, we move e_2 aside. Next we shuttle electron e_1 to n_2 in order for A-gate S_2 above n_2 to apply hyperfine interaction, as shown in Fig. 2.13. Through this interaction, the two qubits become entangled. We would like to have pure hyperfine evolution according to the H_A part of Eq. (2.42); however, the magnetic field (the H_B part of Eq. (2.42)), whose role is to augment the interaction, is on all the time. The H_A part should now contain terms that describe the entanglements, in particular $\boldsymbol{\sigma}^{1e}\cdot\boldsymbol{\sigma}^{2N}$ and $\boldsymbol{\sigma}^{2e}\cdot\boldsymbol{\sigma}^{1N}$:

$$H_A = \sum_{i,j} A_{ij}\boldsymbol{\sigma}^{ie}\cdot\boldsymbol{\sigma}^{jN}. \tag{2.48}$$

We can achieve this outcome by "slicing" the time needed for the hyperfine evolution into many steps using the Trotter formula [Skinner et al., 2003; Nielsen and Chuang, 2000]:

$$e^{-iH_A/\hbar} \approx \left[e^{iH_B\Delta t/2\hbar} e^{-i(H_A+H_B)\Delta t/\hbar} e^{iH_B\Delta t/2\hbar} \right]^a . \qquad (2.49)$$

In this way we apply the full Hamiltonian evolution and still have only H_A, i.e., pure hyperfine evolution. To apply this result, we take the number a of $\Delta t = t/a$ steps. The middle term on the right-hand side of Eq. (2.49) corresponds to the full Hamiltonian evolution and the two other terms correspond to a magnetic interaction that corrects the magnetic evolution. The full period of the hyperfine interaction is given as $T_B = h/\Delta E_r$ (see Fig. 2.12). The hyperfine period is $T_A = h/4A$.

Using digital bit trains that apply voltage pulses to A-gates, we can construct arbitrary quantum logic gates with the help of ϕ pulses of magnetic evolution, (B, ϕ), and θ pulses of pure hyperfine evolution, (A, θ), where

$$(B, \phi) \equiv e^{-iH_B\phi T_B/h}, \qquad (A, \theta) \equiv e^{-iH_A\theta T_A/h}. \qquad (2.50)$$

For instance, CNOT can be implemented as follows:

$$\text{CNOT} = M\,N\,M^\dagger, \qquad (2.51)$$

where M contains only single qubit operations

$$M = \left(B, \frac{3\pi}{2}\right)(A_{11} + A_{22}, \pi)\left(A_{11}, \frac{\pi}{2}\right)\left(B, \frac{\pi}{2}\right),$$

and N contains only mixed qubit operations that stem from the entanglement

$$N = \left(A_{12} + A_{21}, \frac{3\pi}{2}\right)\left(B, \frac{\pi}{2}\right)(A_{21}, \pi)\left(B, \frac{3\pi}{2}\right)\left(A_{12} + A_{21}, \frac{\pi}{2}\right).$$

This outcome could be simplified if we used two different g factors, g_1 and g_2 [Levy, 2002] (neighboring ^{31}P's could be embedded into different surrounding substrates). But for the time being, this result, is only a theoretical, because placing ^{31}P's into just one medium—that would be pure enough—and into precisely defined locations is already an extremely demanding task.

Stressing that the Kane computer—as we have seen above—satisfies the five requirements for the implementation of quantum computation given in Sec. 2.6, we can conclude that the Kane computer is a promising technological candidate for a would-be quantum computer. Although the challenges facing a realization of this project are tremendous, it is expected that the current advances in "shrinking" conventional silicon electronics will inevitably lead to applicable new solutions [Clark et al., 2003; Schenkel et al., 2003].

2.5 Ion Traps

Another promising candidate for a possible would-be quantum computer makes use of cold trapped ions [Cirac and Zoller, 1995]. In the last ten years, experiments with trapped ions have yielded results competitive with those achieved on NMR setups. Besides, these experiments can be modified so as to enable scalability [Cirac and Zoller, 2000; Cirac and Zoller, 2004]. Realistic implementations of quantum calculation with ions face tremendous problems, but the physics behind the process has been worked out more completely than the physics of the Kane computer, where there are still unanswered questions.

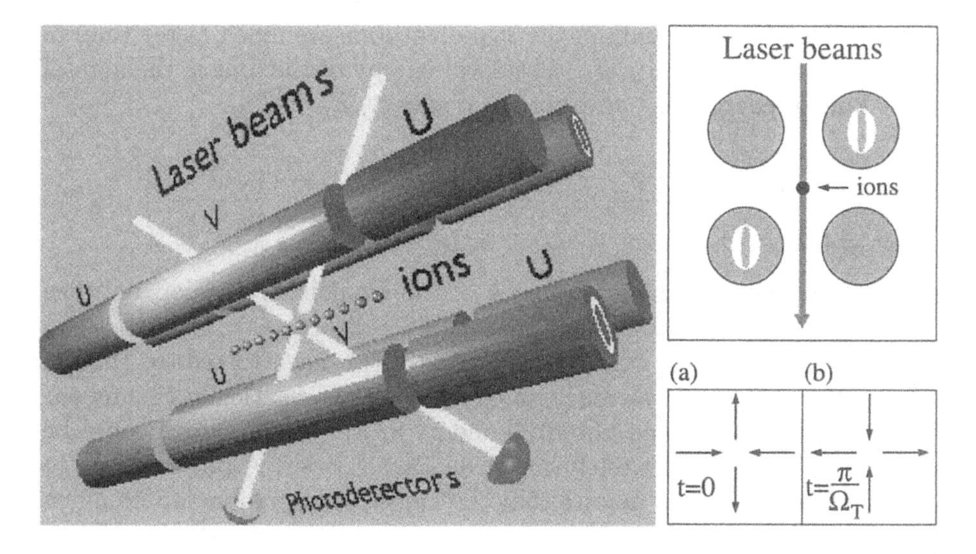

Figure 2.14. A possible ion trap realization for a large number of ions: Paul's linear trap (not to scale) [Cirac and Zoller, 1995; Steane, 1997]. The opposite end sections of two of the rods are under static positive voltage U to force the ions to stay at a constant average distance from each other. An AC voltage $V = V_0 \cos(\Omega_T t)$ is applied to the middle parts of these rods. The other two rods are grounded (0). The insets (a) and (b) show the dynamics of the electric fields at $t = 0$ and $t = T/2$, respectively, where T is the period of the AC current.

Ion traps used in experiments are mostly Paul traps (electrical field only, as opposed to Penning traps, which use both electric and magnetic fields). To confine ions to well-localised positions in space, we must use a time-dependent RF (see p. 96) electric field.[3] To trap large numbers of ions, we cannot use the standard spherical traps with hyperbolic fields, which have an exact solution [Demtröder, 1996, 14.2.1], because in such a trap we cannot keep all the ions in the middle of the trap where the RF field is zero. This would cause too large a motion and therefore the heating of ions.

The so-called *linear ion trap*, shown in Fig. 2.14, is used for trapping large numbers of ions. The RF field is zero along the central line of the electrode configuration. The ions are confined along this line with static electric fields: the ends of the segmented rods are held at a positive potential [Steane, 1997]; alternatively, one can have a static ring around the end sections of the AC electrodes or just have positively charged points at the each end of the ion line. Let this line be our x-axis. If the potentials are chosen appropriately, cold ions may be harmonically trapped in all three dimensions. The directions of the corresponding fields, shown in insets (a) and (b) of Fig. 2.14, determine the x- and y-axes. The displacements in the x, y-directions are much faster than in the z-direction: $\omega_z \ll \omega_x, \omega_y$. The total energy of the ions is the sum of their kinetic energy, E_k, and potential energy, E_p:

$$E_k = \sum_{i=1}^{N} \frac{p_{ix}^2 + p_{iy}^2 + p_{iz}^2}{2m}, \qquad E_p = \sum_{i=1}^{N} \left(\zeta z_i^n + \sum_{i>j}^{N} \frac{Q^2}{|z_i - z_j|} \right), \tag{2.52}$$

where the potential energy consists of the Coulomb repulsion between ions and a term describing trapping in the z-direction.

To get ion behavior that we can control, we have to reduce the energy in the z direction as much as possible—mathematically this means obtaining a well-behaved potential energy in Eq. (2.52), which could be approximated by the lowest terms of its Taylor series. A physical way to achieve this result is to require that the kinetic energy of an ion be much less then the quantum of energy corresponding to its vibration in the z-direction, $\hbar\omega_z$. Such vibrational quanta of energy are called *phonons*. Since thermodynamically the kinetic energy of particles corresponds to their temperature, this requirement can be expressed as $k_B T \ll \hbar\omega_z$, where k_B is the Boltzmann constant and T is the temperature; i.e., we

[3] Basic electrostatics tells us that it is not possible to design an ion trap using only static fields. More precisely, Gauss' Law (1st Maxwell's equation) tells us that in a charge-free region of space, there can be no local minimum (or maximum) in the potential. This result can also be expressed by the continuity equation: the divergence of the electric field is equal to zero—hence all the field lines going to the center of the ion trap must come out of the trap center in some other direction. Hence a positive charge cannot be confined there. This result is sometimes referred to as *Earnshaw's theorem* (Earnshaw first arrived at it in 1842).

have to cool down the ions. However, here we have an example of microscopic cooling in one dimension, while in the other two the ions oscillate rapidly ($\omega_z \ll \omega_x, \omega_y$).

Therefore, the cooling cannot be an overall thermodynamical cooling but has to be a targeted reduction of the kinetic energy of ions in a chosen direction. This reduction can be accomplished by Doppler laser cooling. The process works as follows. First, we need a near-perfect vacuum so that there are no other atoms apart from our ions that could possibly kick the ions. Then we direct laser beams along the z-axis towards ions in the trap. Only photons of a frequency that can excite an atom (ion) will interact with it. Photons of other frequencies cannot "see" the atom because they have wavelengths that are much longer than the dimensions of atoms. We also say that the atom has a much bigger cross section for the former photons. Now an atom moving towards a laser beam has higher transition frequencies than one moving away from it (Doppler effect), and if we tune the laser beam so as to match the former frequencies, the photons will kick (transfer their momentum $p = h/\lambda$ to) the atoms coming towards them and will not interact with the atoms going away from them. We also must apply so-called *Sisyphus cooling* (see p. 119) until we reach the temperature $T \ll \hbar\omega_z/k_B$.

For ions with energy reduced in this way in the z-direction (we also limit the kinetic part to the z-axis), we can expand the potential energy in Eq. (2.52) around the equilibrium position and approximate it by only the first terms of the Taylor series. We get quadratic terms (for $j = i$) and mixed terms (for $j \neq i$), but by expressing these terms with the help of normal modes of oscillation (see Fig. 2.15 (b)), we get rid of mixed terms and obtain the harmonic oscillator Hamiltonian

$$H = \sum_{i=1}^{N} \left(\frac{p_{zi}^2}{2m} + \frac{1}{2} m\omega_z^2 z_i^2 \right), \tag{2.53}$$

where H, p_{zi}, and z_i are operators, but we drop the "hats" to ease the notation. The last two operators satisfy the commutation rules:

$$[z_i, p_{zi}] = i\hbar. \tag{2.54}$$

In Sec. 1.17 (for Eq. (1.69)), we introduced the annihilation operator a and several "rules of thumb" for applying it to state vectors. Its Hermitian conjugate, $a^\dagger$, is called the *creation operator* and $N = a^\dagger a$ is the *number operator*. These names are clarified by their actions on states vectors given below.

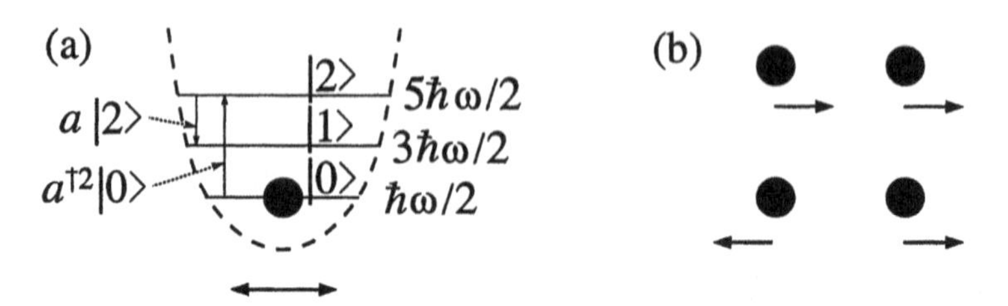

Figure 2.15. (a) Harmonic oscillator potential; (b) normal modes: the upper mode is the left-to-right motion of both ions together, i.e., the center of mass (CM) mode, and the lower one is the one in which CM does not move, the so-called *stretch mode.*

We define the *annihilation operator* for each ion as follows:

$$a = \sqrt{\frac{1}{2\hbar}}\left(\sqrt{m\omega_z}z + \frac{ip_z}{\sqrt{m\omega_z}}\right). \tag{2.55}$$

Using the corresponding creation and number operators and the commutation rule given by Eq. 2.54 we can write down the Hamiltonian (2.53) applied to its eigenstates $|n\rangle$ as

$$H|n\rangle = \frac{\hbar\omega_z}{2}(aa^\dagger + a^\dagger a)|n\rangle = \hbar\omega_z(N + \frac{1}{2})|n\rangle = E_n|n\rangle, \tag{2.56}$$

We emphasize here that $|n\rangle$ means an n-phonon state. Thus, $|0\rangle$ and $|1\rangle$ mean states containing zero and one phonons.

Note that phonons are bosons and that a and $a^\dagger$ satisfy the following anticommutation relations:

$$a_j^\dagger a_k + a_k a_j^\dagger = \delta_{jk}. \tag{2.57}$$

The names *creation* and *annihilation operators* come from the following relations one can easily derive [Messiah, 1965]:

$$a^\dagger|n\rangle = \sqrt{n+1}|n+1\rangle, \quad a|n\rangle = \sqrt{n}|n-1\rangle \quad n \neq 0, \quad a|0\rangle = 0.$$

We also obtain $N|n\rangle = n|n\rangle$, which clarifies the name number operator, as well as

$$E_0 = \frac{1}{2}\hbar\omega_z, \quad E_1 = \frac{3}{2}\hbar\omega_z, \quad \dots \quad E_n = (n + \frac{1}{2})\hbar\omega_z, \quad \dots, \tag{2.58}$$

shown in Fig. 2.15.

Therefore the ground state $|0\rangle$ corresponds to no phonons, i.e., to the CM motion with frequency ω_z, and has the energy $\hbar\omega_z/2$ for each phonon. The first mode corresponds to the number state 1 and to one phonon, and since the ions are moving in opposite directions, their frequency is (cf. Fig. 2.15) $\sqrt{3}\omega_z$, etc. It can be shown that these frequencies do not change when we increase the number of ions, and this finding is important for the scalability of the setup.

To obtain qubits for computation, we combine these global, phonon states with individual states of particular ions. The latter states we obtain by applying laser beams to individual ions. These states are two-level states that we can obtain in several ways: (a) by Zeeman splitting (Sec. 2.2, 88), achieved with the help of a magnetic field applied to the ground state, (b) by the two-beam Raman scheme for resolving sublevels of a ground state of ions with hyperfine interaction, and (c) by optical transitions between fine structure states for ions with zero nuclear angular momentum, etc. The experimental realizations are too numerous [Wineland et al., 1998; Kielpinski et al., 2002; Cirac and Zoller, 2000; Cirac and Zoller, 2004] to be reviewed here. Therefore, we shall present a schematic of one of the first proposals that remains a very instructive way of combining individual and collective states [Cirac and Zoller, 1995].

Before we dwell on the proposal itself, let us first consider a semiclassical description of a laser controlling a two-level system, as shown in Fig. 2.16. The laser beam of frequency ω_L induces the transitions between the ground and excited levels of an atom. We describe the beam by the classical electromagnetic plane wave given by Eq. (1.8). The equation's real part is

$$\mathbf{E} = \mathbf{E}_0 \cos(\mathbf{k} \cdot \mathbf{r} - \omega t + \phi), \tag{2.59}$$

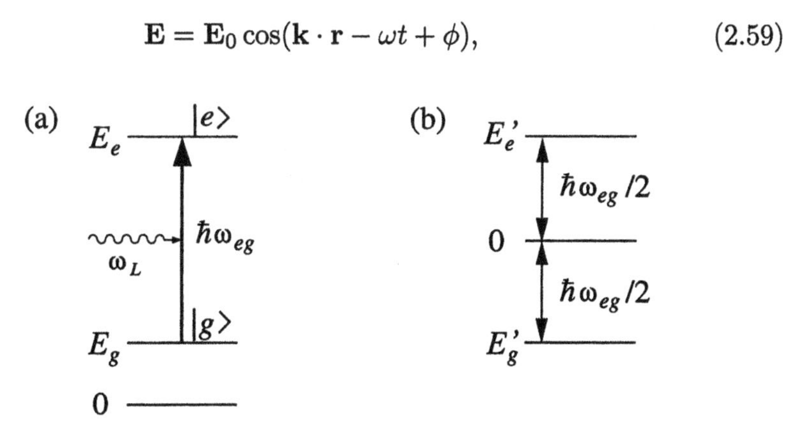

Figure 2.16. (a) A two-level system interacting with a laser beam field; (b) redefined zero-energy—see Eq. (2.87).

Since the wavelength of the laser beam is large compared to the size of an atom, we can neglect $\mathbf{k} \cdot \mathbf{r} = 2\pi r/\lambda \ll 1$ for r less than or equal the atom size. This approximation, when connected with a dipole-field interaction (see Eq. (2.61)), is called the *dipole approximation.* We can also drop the phase shift ϕ, since it will make no difference to the final result, which consists in determining the probabilities of electrons occupying the two possible levels. We shall use only the real part of $\mathbf{E}$ from Eq. (1.8):

$$\mathbf{E} = \mathbf{E}_0 \cos \omega t = \frac{\mathbf{E}_0}{2}(e^{i\omega t} + e^{-i\omega t}). \tag{2.60}$$

Hence, our Hamiltonian

$$H = H_0 + \boldsymbol{\mu} \cdot \mathbf{E} \tag{2.61}$$

in the dipole approximation reads

$$H = H_0 - e\mathbf{r} \cdot \mathbf{E}_0 \cos \omega t. \tag{2.62}$$

It determines the Schrödinger equation

$$H\Psi = i\hbar \frac{\partial \Psi}{\partial t} \tag{2.63}$$

whose general solution for our two-level system is

$$|\Psi\rangle = c_g(t) e^{-iE_g t/\hbar} |g\rangle + c_e(t) e^{-iE_e t/\hbar} |e\rangle, \tag{2.64}$$

where $|g\rangle$ and $|e\rangle$ are the base states (ground and excited). Taking into account that

$$H_0|g\rangle = E_g|g\rangle \qquad \text{and} \qquad H_0|e\rangle = E_e|e\rangle \tag{2.65}$$

by introducing Eq. (2.64) into Eq. (2.63) and carrying out the spatial integration, we get [Demtröder, 1996, 2.6.2]

$$\frac{dc_g(t)}{dt} = \frac{iR_{eg}}{2} \left[e^{-i(\omega_{eg}+\omega_L)t} + e^{-i(\omega_{eg}-\omega_L)t} \right] c_e(t), \tag{2.66}$$

$$\frac{dc_e(t)}{dt} = \frac{iR_{eg}}{2} \left[e^{i(\omega_{eg}+\omega_L)t} + e^{i(\omega_{eg}-\omega_L)t} \right] c_g(t), \tag{2.67}$$

where $\omega_{eg} = (E_e - E_g)/\hbar$ and $R_{eg} = D_{eg}E_0/\hbar = R_{eg}$, where D_{eg}, a spatial integral of $\langle g|\mathbf{p} \cdot \mathbf{E}|e\rangle$, is called the *atomic dipole matrix element*—it is determined by the charge distribution in the states $|g\rangle$ and $|e\rangle$. We see that, on the right side of Eq. (2.66), we have only $c_e(t)$, and in Eq. (2.67) only $c_g(t)$. This is because $D_{gg} = D_{gg} = 0$. Physically, it means that $|e\rangle$ can only evolve from $|g\rangle$, and $|g\rangle$ only from $|e\rangle$.

For $\omega_L \approx \omega_{eg}$, the terms containing $\exp[\mp i(\omega_{eg} + \omega_L)t]$ in Eqs. (2.67) and (2.66) oscillate rapidly in time and can be neglected with respect to the near resonant terms, i.e., the terms containing $\exp[\mp i(\omega_{eg} - \omega_L)t]$. This approximation is called the *rotating wave approximation*, in which Eqs. (2.67) read [Demtröder, 1996, 2.6.6]

$$\begin{aligned}\frac{dc_g(t)}{dt} &= \frac{iR_{eg}}{2}e^{-i(\omega_{eg}-\omega_L)t}c_e(t),\\ \frac{dc_e(t)}{dt} &= \frac{iR_{eg}}{2}e^{i(\omega_{eg}-\omega_L)t}c_g(t).\end{aligned} \tag{2.68}$$

Their solutions are

$$\begin{aligned}c_g(t) &= \frac{1}{\mu_1-\mu_2}\left(\mu_1 e^{i\mu_2 t} - \mu_2 e^{i\mu_1 t}\right),\\ c_e(t) &= \frac{iR_{eg}}{\Omega}e^{i(\omega_{eg}-\omega_L)t/2}\sin\frac{\Omega t}{2},\end{aligned} \tag{2.69}$$

where

$$\begin{aligned}\mu_{1,2} &= -\frac{1}{2}(\omega_{eg}-\omega_L) \pm \sqrt{(\omega_{eg}-\omega_L)^2 + R_{eg}^2},\\ \Omega = \mu_1 - \mu_2 &= \sqrt{(\omega_{eg}-\omega_L)^2 + R_{eg}^2}.\end{aligned} \tag{2.70}$$

Eqs. (2.69) gives the transition probabilities

$$|c_e(t)|^2 = \left(\frac{R_{eg}}{\Omega}\right)^2 \sin^2\frac{\Omega t}{2}, \qquad |c_g(t)|^2 = 1 - |c_e(t)|^2, \tag{2.71}$$

which oscillate with frequency Ω—called the *Rabi flopping frequency*—between levels E_g and E_e.

For instance, at resonance ($\omega_L = \omega_{eg}$) after a time $T = \pi/\Omega = \pi/R_{eg}$ the probability of finding the system in level E_e is 1. This result means we will have the following evolution:

$$\begin{aligned}|c_g(0)|^2 = 1 &\to |c_g(T)|^2 = 0, \qquad \text{and}\\ |c_e(0)|^2 = 0 &\to |c_e(T)|^2 = 1,\end{aligned} \tag{2.72}$$

i.e., full control over the states of the electron in time.

We note here that, in the literature, R_{eg} is often called the *Rabi frequency* and denoted Ω_{eg} whenever $\omega_L = \omega_{eg}$ or when only $\omega_L \approx \omega_{eg}$ because Eq. (2.70) then reads $\Omega = R_{eg}$ and $\Omega \approx R_{eg}$, respectively. For example, Eq. (2.68) would in this notation read[4]

$$\begin{bmatrix} \frac{dc_g}{dt} \\ \frac{dc_e}{dt} \end{bmatrix} = \frac{i\,\Omega_{eg}(t)}{2} \begin{bmatrix} 0 & e^{-(i\omega_{eg}-\omega_L)t} \\ e^{i(\omega_{eg}-\omega_L)t} & 0 \end{bmatrix} \begin{bmatrix} c_g \\ c_e \end{bmatrix}. \tag{2.73}$$

So far, we have used a semiclassical description of the interaction between a two-level atom and a laser field. A quantum description of the interaction would require a quantization of the radiation field and dealing with its Fock states, i.e., photon number states. This description is known as the *Jaynes–Cummings model*. However, since the model essentially serves as a bridge to a model describing the interaction of a laser field and collective modes of trapped ions—a formal similarity between photon and phonon behavior has been recognized [Cirac et al., 1993]—we can dwell directly on the laser beam–electron–phonon interaction as presented in [Cirac and Zoller, 1995].

To this end, we shall repeat the above procedure, starting again with the electric field given by Eq. (2.59) and the Hamiltonian (2.61). However, we shall not use the dipole approximation (Eqs. (2.61) and (2.62)), because a laser beam that acts on ions should not only induce transitions between their ground and excited levels but also change the states of their collective modes. Since the direction and amount of the wave vector $\mathbf{k}$ of a laser beam must be taken into account to describe its interaction, the term $\mathbf{k}\cdot\mathbf{r}$ cannot be neglected as in the dipole approximation. Instead, we start with [Sørensen and Mølmer, 1999; Sørensen and Mølmer, 2000]

$$H = H_0 + H_{\text{int}} = H_{\text{col}} + H_{\text{el}} + H_{\text{int}}, \tag{2.74}$$

where H_{col} is the Hamiltonian of the collective ion mode given by Eq. (2.56); the laser beam–ion interaction term H_{int} is

$$H_{\text{int}} = \boldsymbol{\mu}\cdot\mathbf{E} \tag{2.75}$$

where $\boldsymbol{\mu}$ is an electric dipole created in the ion; and H_{el} can be expressed as

$$H_{\text{el}} = \sum_j E_j \varepsilon_j^\dagger \varepsilon_j, \tag{2.76}$$

[4]We are going to refer to Eqs. (2.68) and (2.73) in Sec. (3.1.4) (p. 142) while deriving an equivalent equation—Eq. (3.4)—for three levels, and the references we are going to cite in Sec. (3.1.4) will make use of the Rabi frequency notation.

where $\varepsilon^\dagger$ and ε are electron creation and annihilation operators, respectively. Note that electrons are fermions and that ε and $\varepsilon^\dagger$ satisfy the following commutation relations (cf. Eq. (2.57)):

$$\varepsilon_j^\dagger \varepsilon_k - \varepsilon_k \varepsilon_j^\dagger = \delta_{jk}. \tag{2.77}$$

To describe the Jaynes–Cummings model, one usually makes use of a formal analogy between a two-level atom and a spin $\frac{1}{2}$ system in a magnetic field. Such an analogy has been used for descriptions of trapped-ion models as well [Cirac et al., 1993]. It consists in calling the electric dipole $\boldsymbol{\mu}$ and the electric field $\mathbf{E}$ from Eq. (2.61) a *fictitious magnetic dipole* $\boldsymbol{\mu}_\sigma$ and a *fictitious magnetic field* $\mathbf{B}_\sigma$, respectively. This enables us to introduce *fictitious spin operators* σ^+, σ^-, and σ_z as follows.

The main aim of such fictitious operators is to simplify the state notation. In the number state notation, we denote the vacuum state by $|0\rangle$ and a state occupied by one electron by $|1\rangle$. So the ground state in a two-level system is denoted by $|10\rangle = |1,0\rangle = |1\rangle|0\rangle$ and the excited state by $|01\rangle = |0,1\rangle = |0\rangle|1\rangle$. However, since in quantum optics we are mainly concerned with single electrons, it is practical to reduce this two-state formalism to a one-state formalism. To do this, let us look at the actions of the *Fermi operators*:

$$\varepsilon_1^\dagger \varepsilon_2|01\rangle = \varepsilon_1^\dagger|00\rangle = |10\rangle \qquad \varepsilon_1^\dagger \varepsilon_2|10\rangle = 0. \tag{2.78}$$

We shall deal neither with $|00\rangle$ nor with $|11\rangle$, but if we did, we would get $\varepsilon_1^\dagger\varepsilon_2|00\rangle = 0$ and $\varepsilon_1^\dagger\varepsilon_2|11\rangle = 0$, respectively. (We have $\varepsilon^\dagger|1\rangle = 0$ because electrons are fermions, and two of them cannot occupy the same state.) Similarly, we have

$$\varepsilon_2^\dagger \varepsilon_1|01\rangle = 0, \qquad \varepsilon_2^\dagger \varepsilon_1|10\rangle = \varepsilon_2^\dagger|00\rangle = |01\rangle, \tag{2.79}$$

$$(\varepsilon_2^\dagger \varepsilon_2 - \varepsilon_1^\dagger \varepsilon_1)|01\rangle = |01\rangle, \qquad (\varepsilon_2^\dagger \varepsilon_2 - \varepsilon_1^\dagger \varepsilon_1)|10\rangle = -|10\rangle. \tag{2.80}$$

If we now introduce (bearing in mind that we shall deal neither with $|00\rangle$ nor with $|11\rangle$)

$$|01\rangle = |g\rangle = \begin{bmatrix} 0 \\ 1 \end{bmatrix} \qquad \text{and} \qquad |10\rangle = |e\rangle = \begin{bmatrix} 1 \\ 0 \end{bmatrix} \tag{2.81}$$

in analogy with Eqs. (2.78) and (2.79), we have

$$\sigma^+|g\rangle = \sigma^+\begin{bmatrix} 0 \\ 1 \end{bmatrix} = \begin{bmatrix} 1 \\ 0 \end{bmatrix} = |e\rangle, \qquad \sigma^+|e\rangle = 0,$$

$$\sigma^-|g\rangle = 0, \qquad \sigma^-|e\rangle = \sigma^-\begin{bmatrix} 1 \\ 0 \end{bmatrix} = \begin{bmatrix} 0 \\ 1 \end{bmatrix} = |g\rangle, \tag{2.82}$$

where

$$\sigma^+ = \frac{1}{2}(\sigma_x + i\sigma_y) = \begin{bmatrix} 0 & 1 \\ 0 & 0 \end{bmatrix}, \quad \sigma^- = \frac{1}{2}(\sigma_x - i\sigma_y) = \begin{bmatrix} 0 & 0 \\ 1 & 0 \end{bmatrix}, \tag{2.83}$$

where σ_x and σ_y are the Pauli matrices defined in Eq. (1.88). Also, in analogy with Eq. (2.80), we have

$$\sigma_z|g\rangle = |g\rangle \qquad \text{and} \qquad \sigma_z|e\rangle = -|e\rangle, \tag{2.84}$$

where the Pauli matrix σ_z is given by Eq. (1.88).

Thus the Hamiltonian H_{el} given by Eq. (2.76),

$$H_{\text{el}} = E_e\varepsilon_2^\dagger\varepsilon_2 + E_g\varepsilon_1^\dagger\varepsilon_1, \tag{2.85}$$

can be written as

$$\begin{aligned} H_{\text{el}} &= \frac{1}{2}E_{eg}(\varepsilon_2^\dagger\varepsilon_2 - \varepsilon_1^\dagger\varepsilon_1) + \frac{1}{2}(E_e + E_g)(\varepsilon_2^\dagger\varepsilon_2 + \varepsilon_1^\dagger\varepsilon_1) \\ &= \frac{1}{2}E_{eg}\sigma_z + \frac{1}{2}(E_e + E_g)\mathbb{1} = \frac{1}{2}\omega_{eg}\hbar\sigma_z + \frac{1}{2}(E_e + E_g)\mathbb{1}, \end{aligned} \tag{2.86}$$

where $E_{eg} = E_e - E_g$. By redefining the zero-energy so as to put it in between E_e and E_g, as shown in Fig. 2.16 (p. 113, we get

$$H_{\text{el}} = \frac{1}{2}\omega_{eg}\hbar\sigma_z, \qquad H_{\text{el}}|g\rangle = -\frac{1}{2}\omega_{eg}|g\rangle, \qquad H_{\text{el}}|e\rangle = \frac{1}{2}\omega_{eg}|e\rangle. \tag{2.87}$$

In a similar way, we can redefine H_{col} from Eq. (2.74) given by Eq. (2.56) so as to suppress the zero-point energy $\hbar\omega_z/2$. Hence we obtain

$$H_0 = H_{\text{col}} + H_{\text{el}} = \hbar\omega_z a^\dagger a + \frac{1}{2}\omega_{eg}\hbar\sigma_z. \tag{2.88}$$

The interaction term H_{int} (2.75) of our Hamiltonian (2.74) is [Wineland et al., 1998; Wineland et al., 2003]

$$H_{\text{int}} = \boldsymbol{\mu}\cdot\mathbf{E} = \mu(\sigma^+ + \sigma^-)E_0\cos(\mathbf{k}\cdot\mathbf{r} - \omega_L t + \phi). \tag{2.89}$$

To simplify the presentation and the equations, we have assumed (without loss of generality) that $\mathbf{r}$ (the ion position operator with respect to the equilibrium position of the ion) is oriented along the axis of the trap and that the wave vector $\mathbf{k}$, i.e., the laser beam, is also oriented along that axis. So we have

$$r = z = z_0(a + a^\dagger), \qquad \text{where} \qquad z_0 = \sqrt{\langle 0|z|0\rangle} = \sqrt{\frac{\hbar}{2m\omega_z}}. \tag{2.90}$$

The electric field is assumed to be polarized along the x-axis, and thus the electric dipole $\boldsymbol{\mu}$ is an operator for the internal transition. Therefore, it is, in analogy with electronic spin (see Eq. (2.6)), proportional to $\sigma_x = \sigma^+ + \sigma^-$; ϕ is the phase of the laser beam at the equilibrium position of the ion, and E_0 is the amplitude of the laser wave.

Taken together, we have

$$H_{\text{int}} = \hbar\Omega(\sigma^+ + \sigma^-)\left\{e^{i[\eta(a+a^\dagger)-\omega_L t+\phi]} + e^{-i[\eta(a+a^\dagger)-\omega_L t+\phi]}\right\}, \quad (2.91)$$

where $\eta = kz_0$ is called the *Lamb–Dicke parameter* and Ω is the Rabi frequency (cf. Eq. (2.71)) and is in our case $\Omega = \mu E_0/(4\hbar)$. (Recall that $\sigma^+ = |e\rangle\langle g|$ and $\sigma^- = |g\rangle\langle e|$.)

Now the laser beam we use for manipulating the ions we also use for cooling the ions further, after the Doppler cooling reaches its limits. A tuned laser beam allows an ion to run up a dipole potential hill (thus losing energy); then, before the ion can run down again and regain the energy, it pumps the ion into another ground state with a lower potential hill. Then the ion returns to the former ground state via another laser pumping, and the whole process starts again. With each cycle, the ion loses some energy, and since it always goes up the potential hill more often than down, the cooling is called *Sisyphus cooling* [Wineland et al., 1992]. To make such cooling possible, the characteristic length scale of motion of the ion must be much smaller than the wavelength of the exciting laser beam. This is called the *Lamb–Dicke limit.* It implies that the frequency of the ion modes (ω_z, $\sqrt{3}\omega_z$, ...) must be larger than the recoil frequency corresponding to the transition used for laser cooling. Actually, the exploration of such cooling processes brought the researchers to the idea of using the same mechanisms for quantum computing, and the teams that were previously engaged in the cooling of atoms were the first to propose and implement ion computing [Wineland and Itano, 1979; Blatt et al., 1986; Wineland et al., 1992; Cirac et al., 1992].

For our purpose of understanding how qubits can be implemented in ion traps, an elaboration of the Hamiltonian (2.91) turns out to be important—namely, its presentation in the interaction picture (in which both the state functions and operators are time dependent) [Greiner, 1989, 10.9]. Starting with the Scrödinger equation

$$(H_0 + H_{\text{int}})\Psi = i\hbar\frac{d\Psi}{dt}, \quad (2.92)$$

we make the transformation

$$\Psi = U\Psi', \qquad \text{where} \qquad U = e^{-iH_0 t/\hbar}, \quad (2.93)$$

and obtain

$$H'_{\text{int}}\Psi' = i\hbar\frac{d\Psi'}{dt}, \qquad \text{where} \qquad H'_{\text{int}} = U^\dagger H_{\text{int}} U. \tag{2.94}$$

To transform the Hamiltonian (2.91) into the interaction picture, we use the following relations between the Schrödinger and interaction picture operator forms:

$$\varepsilon'_1 = \varepsilon_1 e^{-i\omega_g t}, \qquad \varepsilon'_2 = \varepsilon_2 e^{-i\omega_e t}, \qquad \text{and} \qquad a' = a e^{-i\omega_z t}. \tag{2.95}$$

Since σ^+ corresponds to $\varepsilon_2^\dagger \varepsilon_1$ and since $\sigma^- = (\sigma^+)^\dagger$ [see Eq. (2.83)], we get

$$\sigma'^+ = \sigma^+ e^{i\omega_{eg}t}, \qquad \sigma'^- = \sigma^- e^{-i\omega_{eg}t}, \qquad \text{and} \qquad a'^\dagger = a^\dagger e^{i\omega_z t}. \tag{2.96}$$

By introducing these operators into Eq. (2.91), we get an expression with the terms containing

$$e^{-i(\omega_{eg}+\omega_L)t} \qquad \text{and} \qquad e^{-i(\omega_{eg}-\omega_L)t}, \tag{2.97}$$

as well as the terms containing their conjugates.

Here, as in Eq. (2.67), we can apply the *rotating wave approximation* for $\omega_L \approx \omega_{eg}$ and neglect the terms containing $\exp[\mp i(\omega_{eg} + \omega_L)t]$, which rapidly oscillates in time with respect to the terms containing near resonant $\exp[\mp i(\omega_{eg} - \omega_L)t]$. Thus Eq. (2.91) yields our Hamiltonian in the interaction picture:

$$H'_{\text{int}} = \hbar\Omega\sigma^+ e^{i[\eta(a e^{-i\omega_z t} + a^\dagger e^{i\omega_z t}) - \delta + \phi]} + \text{H.c.}, \tag{2.98}$$

where $\delta = \omega_L - \omega_{eg}$ and H.c. means the *Hermitian conjugate.* The corresponding wave function for the first two collective modes is

$$\Psi' = C_{g,0}(t)|g\rangle|0\rangle + C_{g,1}(t)|g\rangle|1\rangle + C_{e,0}(t)|e\rangle|0\rangle + C_{e,1}(t)|e\rangle|1\rangle. \tag{2.99}$$

Of primary interest for quantum computing will be the resonant transitions $\delta = (n' - n)\omega_z$, where $n, n' = 0, 1$.

By introducing Ψ' from Eq. (2.99) into Eq. (2.94), we get, in fashion similar to Eq. (2.67) [Wineland et al., 1998],

$$\begin{aligned} \frac{dC_{e,n'}}{dt} &= -i^{1+|n'-n|} e^{i\phi}\Omega_{n',n} C_{g,n}, \\ \frac{dC_{g,n'}}{dt} &= -i^{1-|n'-n|} e^{-i\phi}\Omega_{n',n} C_{e,n'}, \end{aligned} \tag{2.100}$$

where

$$\Omega_{n',n} = \Omega|\langle n'|e^{i\eta(a+a^\dagger)}|n\rangle| = \Omega e^{-\eta^2/2}\sqrt{\frac{n_<!}{n_>!}}\eta^{|n'-n|} L_{n_<}^{|n'-n|}(\eta^2), \tag{2.101}$$

where $n_>$ ($n_<$) is the greater (smaller) of n' and n, and L_n^α is the generalized Laguerre polynomial

$$L_n^\alpha(X) = \sum_{m=0}^{n}(-1)^m \binom{n+\alpha}{n-m} \frac{X^m}{m!} \,. \tag{2.102}$$

In ion quantum computing experiments, we are primarily interested in $n' = n, n \mp 1$ (in particular, in $n = 0, 1$). The corresponding $\omega_L = \omega_{eg}$, $\omega_{eg} \mp \omega_z$ are called the *carrier, red sideband,* and *blue sideband* frequencies, respectively.

Upon solving Eq. (2.100), we get [Wineland et al., 1998]

$$\begin{aligned} |n\rangle|g\rangle &\to \cos\Omega_{n',n}t|n\rangle|g\rangle - ie^{i(\phi+\frac{\pi}{2}|n'-n|)}\sin\Omega_{n',n}t|n'\rangle|e\rangle, \\ |n\rangle|e\rangle &\to -ie^{i(\phi+\frac{\pi}{2}|n'-n|)}\sin\Omega_{n',n}t|n\rangle|g\rangle + \cos\Omega_{n',n}t|n'\rangle|e\rangle. \end{aligned} \tag{2.103}$$

For the carrier frequency ($n' = n$), we can write Eq. (2.103) as

$$\begin{aligned} |n\rangle|g\rangle &\to \cos\Omega_{n,n}t|n\rangle|g\rangle - ie^{i\phi}\sin\Omega_{n,n}t|n\rangle|e\rangle \\ |n\rangle|e\rangle &\to -ie^{i\phi}\sin\Omega_{n,n}t|n\rangle|g\rangle + \cos\Omega_{n,n}t|n\rangle|e\rangle, \end{aligned} \tag{2.104}$$

where $\Omega_{n,n}$, for $n = 0, 1$, we are going to use, follow from Eq. (2.101):

$$\Omega_{0,0} = \Omega e^{-\eta^2/2}, \qquad \Omega_{1,1} = \Omega e^{-\eta^2/2}(1-\eta^2). \tag{2.105}$$

Therefore, with a proper choice of laser pulse duration and phase, we can use $|g\rangle$ and $|e\rangle$ as single qubit states to set up a single qubit gate (see Sec. 1.18, p. 45). For example, for a laser pulse of duration $t = \pi/\Omega_{n,n}$ and a phase $\phi = -\pi/2$, we get the NOT gate (1.32). On the other hand, $t = \pi/(2\Omega_{n,n})$, $\phi = -\pi/2$, and a rotation about the z-axis give a Hadamard gate (1.59).

To construct a CNOT gate, we need a two-qubit system, and for that purpose we can use a single ion with collective states $|0\rangle$ and $|1\rangle$ as the control qubit states and $|g\rangle$ and $|e\rangle$ as the target qubit states. Thus we will have

$$|00\rangle = |0\rangle|g\rangle, \quad |01\rangle = |0\rangle|e\rangle, \quad |10\rangle = |1\rangle|g\rangle, \quad |11\rangle = |1\rangle|e\rangle. \tag{2.106}$$

Our aim is to swap $|g\rangle$ and $|e\rangle$ whenever the ion is in state $|1\rangle$ and to leave them unchanged whenever it is in $|0\rangle$—by a single laser pulse. To achieve this aim, Monroe, Leibfried, King, Meekhof, Itano, and Wineland [Monroe et al., 1997] set the Lamb–Dicke parameter η so that[5]

$$\frac{\Omega_{1,1}}{\Omega_{0,0}} = \frac{2k+1}{2m}. \tag{2.107}$$

[5]Alternative CNOT implementations have been put forward by Cirac and Zoller [Cirac and Zoller, 1995; Poyatos et al., 2000].

This can be done by choosing (see Eq. (2.105))

$$\eta = \sqrt{1 - \frac{2k+1}{2m}}. \tag{2.108}$$

Now, driving the laser carrier transition for a duration of $t = (k + 1/2)/\Omega_{1,1}$ is equivalent to driving it for $t = (2m)/\Omega_{0,0}$. Hence, Eq. (2.104) for $k = 0$, $m = 1$, ($\eta = 0.707$), and $\phi = -\pi/2$ yields

$$\begin{aligned}
|00\rangle &\rightarrow \cos 2\pi|0\rangle|g\rangle - ie^{-i\pi/2}\sin 2\pi|0\rangle|e\rangle = |00\rangle \\
|01\rangle &\rightarrow -ie^{-i\pi/2}\sin 2\pi|0\rangle|g\rangle + \cos 2\pi|0\rangle|e\rangle = |01\rangle, \\
|10\rangle &\rightarrow \cos\frac{\pi}{2}|1\rangle|g\rangle - ie^{-i\pi/2}\sin\frac{\pi}{2}|1\rangle|e\rangle = |11\rangle \\
|11\rangle &\rightarrow -ie^{i\phi}\sin\frac{\pi}{2}|1\rangle|g\rangle + \cos\frac{\pi}{2}t|1\rangle|e\rangle = |10\rangle,
\end{aligned} \tag{2.109}$$

which is nothing but a CNOT gate.

To read out the result of a completed calculation the states of all qubits must be measured. For example, let the $|0\rangle$ state of a qubit be a ground state of an ion and $|1\rangle$ a metastable excited state of the ion, as in Fig. 3.3 (p. 140). We can interrogate the qubit with a laser beam tuned to excite the electron from the ground state to an excited state (the transition at 397 nm in Fig. 3.3, p. 140). If the ion then emits a photon, this means that the ion was in the ground state. If it does not emit a photon, it is in the metastable level (the laser beam does not "see" the ion; it can be seen there only by a laser tuned to a transition from the metastable to the excited level—866 nm).

In Sec. 1.22 (p. 72), we have seen that by using single qubit and CNOT gate we can entangle qubits. However, we have also seen (in Sec. 1.20, p. 56) that entangling photons requires selecting appropriate states and throwing away the remaining events. An ion computer can do the entanglement better. It can *deterministically* entangle states *on demand* without throwing away any events [Kielpinski et al., 2002]. An immediate consequence is deterministic quantum teleportation of qubits around ion computer circuits and nets [Barrett et al., 2004].

Since any quantum circuit can be constructed by means of single qubit gates and CNOT gates, the ion computer proves to be universal, and since any two ion states can be entangled no matter how far away from each other they are in the trap, the ion computer is scalable in principle. There are several proposals on how to scale up to realistic ion computers. One way is to store ions in an array of microtraps, which can be realized by electric and/or laser fields as shown in Fig. 2.17 [Cirac and Zoller, 2000]. The key idea is to address neighboring ions conditionally, based on their state, as opposed to the exchange of phonons corresponding to

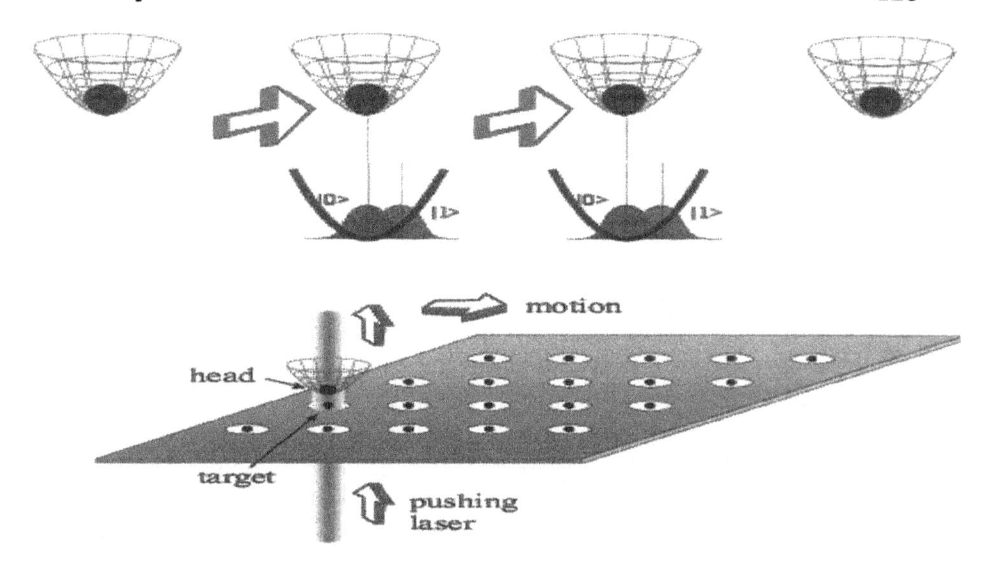

Figure 2.17. A scale-up proposal for ion computers by means of arrays of microtraps, which can be realized by electric and/or laser fields [Cirac and Zoller, 2000].

the collective center-of-mass motion of the ions in the model presented above. Another model is a *quantum charge-coupled device* architecture consisting of a large number of interconnected ion traps in each of which one can manipulate a few ions using the methods demonstrated above [Kielpinski et al., 2002]. All in all, these models are technological challenges for future development.

2.6 Future Experiments

There are five generally accepted requirements for the implementation of quantum computation:

1. A scalable physical system with well-characterized qubits,
2. The ability to initialize the state of the qubits to a simple reliable state ($|000...\rangle$),
3. Long decoherence times, much longer than the gate operation time,
4. A universal set of quantum gates,
5. A qubit-specific measurement capability,

and two additional networkability conditions:

6. The ability to interconvert stationary and flying qubits,
7. The ability to faithfully transmit flying qubits between specified locations.

These requirements are called the *DiVincenzo Criteria* [DiVincenzo, 2000].

We have seen that the trapped-ion computer satisfies all these criteria, while the Kane solid state computer satisfies the first five—for the sixth and seventh, it is too early to decide. A number candidates that satisfy most of these criteria, and there are many more candidates that only partially (at least for the time being) satisfy them (see Table 2.1). Consequently, we cannot have a "winner" at the present stage of research. The challenges facing the realization of this project are tremendous, but still it is expected that the current experimental advances will lead us to a winner eventually. A better understanding of the quantum behavior of quantum systems is needed before promising candidates can be engineered to a larger scale. To help scientists and students on the road to a realistic quantum computer and to enable an informed comparison of different projects, a panel of experts in experimental implementations of quantum computing systems has compiled a report under the name *A Quantum Information Science and Technology Roadmap: Part 1. Quantum Computation* [Hughes et al., 2004]. In the report, all presently pursued quantum computing systems have been compared, results reviewed, and references given.

Table 2.1. Promising criteria for quantum computation candidates according to [Hughes et al., 2004]. "+" means *a potentially viable approach with sufficient proof of principle*, "±" means *a potentially viable approach without sufficient proof of principle*, and "−" means *no viable approach is known.*

	DiVincenzo Criteria						
Candidates	1	2	3	4	5	6	7
NMR	−	±	±	+	±	−	−
Solid State	±	±	±	±	±	−	−
Trapped Ion	±	+	±	+	+	±	±
Cavity QED	±	+	±	±	+	±	±
Neutral Atom	±	+	±	±	±	±	±
Optical	±	±	+	±	±	±	+
Superconducting	±	+	±	±	±	−	−
"Unique" Qubits	a variety of criteria evaluation						

Altogether there are over hundred different systems and subsystems used for implementing qubits by almost as many teams all over the world. Consequently, we are currently gaining an abundance of new results on the engineering of quantum systems together with more and more functional quantum computation systems. At the present level of the experimental effort all over the world, a quantum computer with up to 50 qubits and with error-correcting code will most likely appear by

2012. Whether we will be able to construct a quantum computer that will surpass the fastest classical computers in the near future remains to be seen. It is estimated that such a computer should have more than 1000 qubits and should be able to carry out more than 10^9 operations [Preskill, 1998], and this project will probably be attempted for only one or two "winners." The history of classical computing indicates that we might have its quantum "super-counterpart" sooner than later.

2.7 Quantum Communication Implementation

In contrast to quantum computation experiments, quantum communication experiments are at the moment split into those that are almost ahead of their theory—quantum cryptography—and those that have barely started and for which the first extensive implementations are expected after 2012—qubit DiVincenzo networkability (see DiVincenzo criteria 6 and 7 in Sec. 2.6, p. 124). Therefore, in this section, we will limit ourselves to a brief quantum cryptography implementation overview.

Quantum cryptography is entering its physical and industrial application stage. The protocols currently being used in applications are essentially those proposed two decades ago (see Secs. 1.23 and 1.23), although there are also new proposals. At least three companies, BBN Technologies (Cambridge, MA, USA), ID Quantique (Geneva, Switzerland) and MagiQ Technologies (New York, NY, USA), have released commercial quantum cryptography (quantum key distribution, QKD) systems, and several others are about to do so. These products are still not cost-effective for commercial applications. They can be regarded as prototypes supported by multimillion dollar government-, publicly, and privately funded projects [Ouellette, 2005].[6]

As the most important project, we single out the World's first quantum network—the DARPA quantum network—fully operational since October 23, 2003 [Elliott et al., 2005]. This is a joint project of BBN Technologies, Harvard University, and Boston university, technologically

[6]The following funding allocations include quantum cryptography: DARPA of the Department of Defense (DoD) (2002–2007) with $50 million (Quantum Information Science and Technology, QuIST program); in Europe, Quantum Information Processing and Communications (5th and 6th EU Framework Programme) (1998–2007), about $80 million; in Japan, Japanese Government Organisations Sponsoring Nanotechnology R&D $2.75 billion/year. Similar funding exist in Australia, Australasia, and Canada, including many private contributions; for example, in 2004 a single private donation (by Ophelia and Mike Lazaridis) for quantum computing research in Canada to a single institution (University of Waterloo) reached $33 million].

implemented by BBN Technologies, and supported by DARPA.[7] The DARPA quantum network spans 29 km and does take an eavesdropper (Eve) into account. Two kinds of error correction schemes are implemented in the network and therefore, at least in principle, the network can be considered unconditionally secure.

Since higher rigorous distance limits are essential for successful implementations of QKD systems, researchers are now mostly focused on physical solutions that can increase this limit. Most applications and experiments use optical fibers to connect Alice to Bob or a common EPR source to Alice and Bob. Free space transmission suffers high losses (the present distance limit is about 25 km), but a possible ground–satellite–ground communication is also a research goal. There are three main problems here: the medium, the source, and the detector. The lowest losses in fibers occur near the 1550 nm wavelength—with about 50% loss after about 15 km—which is why these wavelengths are widely used by the telecommunications industry. Unfortunately, the best commercially available single-photon counters operate near 800nm. Thus, better 1550 nm would be desirable. The photon source is a particularly difficult problem. and we consider it next. The potential and status of various proposed implementations were presented in the report *Quantum Information Science and Technology Roadmap: Part 1. Quantum Cryptography* [Bennett et al., 2004], compiled by a panel of quantum cryptography technology experts and shown in Table 2.2.

Table 2.2. Attributes of quantum cryptography implementations: 1. Theoretical security status; 2. Distance limit potential; 3. Speed (bit/sec) potential; 4. Maturity (application readiness); 5. Robustness. Scores given: '+' *high*, '±' *medium*, '−' *low*. According to [Bennett et al., 2004].

	Attributes				
Implementations	1	2	3	4	5
Weak laser pulses	±	+	+	±	±
Single-photon source	+	+	±	−	±
Entangled pairs	+	+	±	±	±
Continuous variables	−	−	+	−	−

Most implementations of quantum cryptography rely on weak laser pulses generated by conventional *diode lasers* over optical fibers because this approach enables the use of present telecommunications technology. We attenuate the laser pulse so that it contains less than one photon in

[7]DARPA is the Defense Advanced Research Projects Agency, the central research and development organization for the Department of Defense (DoD) of USA.

a time window. Most experiments so far used a mean photon number $\mu = 0.1$. The probability of detecting n photons in a pulse, under the assumption that the detections are statistically independent, is given by the *Poisson distribution*, which applies when we approximate the single-photon Fock state by a coherent photon state with a very low μ [Gisin et al., 2002]

$$P(n, \mu) = \frac{\mu^n}{n!} e^{-\mu}. \tag{2.110}$$

Therefore, the conditional probability that a nonempty coherent laser pulse contains more than one photon is given by

$$P(n > 1 | n > 0, \mu) = \frac{1 - P(0, \mu) - P(1, \mu)}{1 - P(0, \mu)} \approx \frac{\mu}{2}. \tag{2.111}$$

Hence, $\mu = 0.1$ means that 5% of nonempty pulses contain more than one photon. Two or more photons per pulse jeopardizes the security of the key transmission the most, since Eve can always split the pulse and let one photon through without changing its state. This approach allows her to learn about the corresponding bit without being caught. We cannot go much below $\mu = 0.1$ with the standard approach, because of the *dark counts* (the detector clicks without photons actually arriving at the detector), This is also an issue with today's detectors. We can hope for a development in detector technology that will decrease detector noise. However, recently a novel approach has been put forward that enables us, in effect, to go five times below $\mu = 0.1$ with existing technology [Hwang, 2003].

The approach consists in catching an eavesdropper (Eve) by means of decoy (fake) pulses that a sender (Alice) sends to a receiver (Bob) in addition to proper signal pulses. The signal and decoy pulses differ only in their photon number distributions (intensities). For instance, suppose $\mu_{\text{decoy}-1} = 0$, $\mu_{\text{decoy}-2} = 0.02$, and $\mu_{\text{signal}} = 0.1$. Eve cannot distinguish a decoy state from a signal state when she splits the pulse. Bob, however, can tell whether Eve has split a decoy pulse or not—the error rates will differ—and if he finds she has, he will abort the transmission. It can be shown that the key generator rate achieves a substantial increase and that the secure QKD is possible over much longer distances than with the standard approach.

Single-photon sources get around the problem of the photon number splitting attack problem since the sources are not probabilistic but deterministic. Ideally, and in principle, they produce n photons "on demand" rather than on average. Such a source is also called a *photon gun*. The first promising experiments on this approach were done in 1998–2000.

They were mostly concentrated on single two-level individual atoms, which would be excited by means of a laser beam of one frequency and subsequently would emit a fluorescence photon at another [Gisin et al., 2002]. The moment of emission can be controlled as shown in Sec. 2.5 [Eqs. (2.70) and (2.71)]. The main problem was the collection efficiency (under 0.1%), since photons from free atoms are in general not emitted in a predetermined direction.

A second approach uses photon emission of electron-hole pairs in a semiconductor quantum dot. An electron-hole pair created in the dot by optical pumping recombines and emits photons at different frequencies, each of which can be distinguished by a filter [Gisin et al., 2002]. Another proposal uses single quantum dots and Weierstrass solid immersion lenses that facilitate light collection [Zwiller et al., 2004].

The third approach is the most demanding but apparently also the most promising. It consists in obtaining fluorescence photons by manipulating trapped atoms or ions and is closely connected to quantum computation with trapped ion systems and cavity QED systems (see Sec 2.5, p. 109). For the most recent papers on all three approaches the reader is directed to the recent *Focus on Single Photons on Demand* [Grangier et al., 2004].

The *entangled pair approach* from Table 2.2[8] aims at increasing the distance limit as well as the security against eavesdropping. In an optical fiber, the probability of absorption of a photon as well as of its depolarization increases exponentially with the length of the fiber. The Malus law (1.9) and the correlation probability for entangled photons pairs (1.71) are of the same functional form (assume that a polarization of one of the photons from the pair has been rotated by 90°) and support the BB84 protocol (see Sec. 1.21, p. 64)—as far as Alice and Bob are concerned—in an identical way. Therefore, by putting a source of entangled singlet-like photon pairs midway between Alice and Bob we can double the distance limit. It would be better that Alice keep the source and that the distance is increased by quantum repeaters (see Sec. 3.1.7, p. 151) because a single photon from a pair does not carry any information and cannot be eavesdropped.

The entangled pair implementations have mostly been carried out by means of parametric down-conversion (the inverse of parametric generation). Parametric down-conversion is a quantum effect in nonlinear optics.

---

[8]For the fourth implementation—continuous variables—we direct the reader to [Bennett et al., 2004].

In Sec. 1.8 (p. 17) we dealt with linear polarization

$$\mathbf{P} = \varepsilon_0 \chi \mathbf{E}, \tag{2.112}$$

where ε_0 is the vacuum permittivity and χ the electric susceptibility. In a nonlinear medium, an intense electric field of one frequency can generate nonlinear polarization at other frequencies. Then the outgoing polarizations $\mathbf{P}=[P_1, P_2, P_3]^T$ become nonlinear with respect to the electric field $\mathbf{E}=[E_1, E_2, E_3]^T$:

$$\mathbf{P} = \varepsilon_0(\chi^{(1)}\mathbf{E} + \chi^{(2)}\mathbf{E}^2 + \ldots), \tag{2.113}$$

where $\chi^{(1)}$ is the first ordered (birefringent) electric susceptibility (a 3×3 tensor), $\chi^{(2)}$ is the second ordered electric susceptibility (media lacking a center of symmetry), etc. The first term in Eq. (2.113) reads

$$\varepsilon_0 \begin{bmatrix} \chi_{11} & \chi_{12} & \chi_{13} \\ \chi_{21} & \chi_{22} & \chi_{23} \\ \chi_{31} & \chi_{32} & \chi_{33} \end{bmatrix} \begin{bmatrix} E_1 \\ E_2 \\ E_3 \end{bmatrix} \tag{2.114}$$

(it describes, for example, the birefringent plates we used in Sec. 1.21, Fig. 1.27), the second term is

$$\varepsilon_0 \begin{bmatrix} \sum_{j,k=1}^{3} \chi_{1jk}^{(2)}(\omega_1, \omega_2, \omega_3) E_j E_k \\ \sum_{j,k=1}^{3} \chi_{2jk}^{(2)}(\omega_1, \omega_2, \omega_3) E_j E_k \\ \sum_{j,k=1}^{3} \chi_{3jk}^{(2)}(\omega_1, \omega_2, \omega_3) E_j E_k \end{bmatrix}, \tag{2.115}$$

where $\chi_{ijk}^{(2)}$ is a second rank tensor with 27 components, etc.

In the parametric down-conversion, one photon of frequency ω_0, called the *pump photon*, is incident on the nonlinear media (birefringent dielectric having the above $\chi^{(2)}$ susceptibility) and polarized along a chosen axis (say $\mathbf{P}_{\text{in}} = [P_1, 0, 0]$, which we will call *extraordinary* polarization). The pump photon breaks up into two photons of lower frequencies, ω_1 and ω_2, called the *signal photon* and the *idler photon*, respectively, where $\omega_1 \geq \omega_2$. The process of conversion is a nonlocal process in the sense that the position within a crystal from where the signal and idler photons emerge is not (and theoretically cannot be) determined. As a consequence, although both signal and idler appear together within femtoseconds, a pinhole that lets one of them through cannot precisely determine the direction of the other (the position of another pinhole which would let the other photon through). For pinholes of equal size this results in efficiency of about 0.05.

Energy ($E = h\nu$) conservation yields

$$E_0 = E_1 + E_2 \qquad \Rightarrow \qquad \omega_0 = \omega_1 + \omega_2, \tag{2.116}$$

while momentum conservation implies the phase-matching condition

$$\mathbf{p}_0 = \mathbf{p}_1 + \mathbf{p}_2 \qquad \Rightarrow \qquad \mathbf{k}_0 = \mathbf{k}_1 + \mathbf{k}_2, \tag{2.117}$$

where $\mathbf{k}$ is the wave vector

$$\mathbf{k} = \frac{n\omega}{c}\mathbf{s}, \tag{2.118}$$

where n is the index of refraction, c the speed of light, and $\mathbf{s}$ a unit vector.

For the uniaxial case (only one optical axis), we distinguish two major types of outgoing photons according to their polarizations:

- *Type-I down-conversion.* The pump photon from a strong laser pump beam is extraordinarily (say, vertically) polarized; the signal and idler photons are ordinarily (horizontally) polarized.

- *Type-II down-conversion* The pump and idler are extraordinarily polarized; the signal photon is ordinarily polarized.

Depending on the crystals we choose, we can obtain the coaxial and intersecting signal-idler cones as shown in Fig. 2.18. Various KDP crystals—for example, AgGaSe2—are used for type-I down-conversion. BBO crystals, beta-barium-borate or β-BaB_2O_4, are typically used for type-II down-conversion.

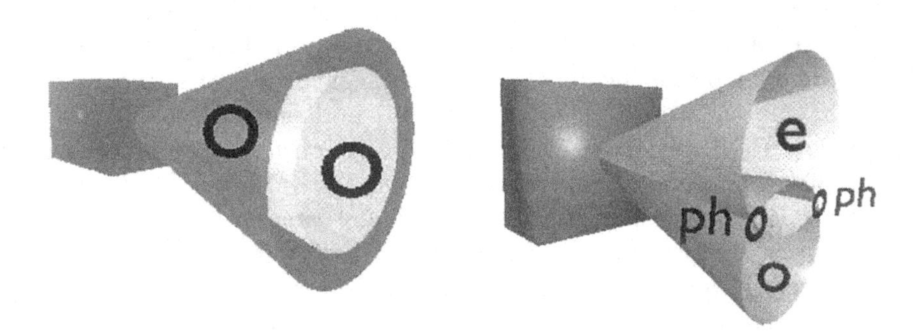

Figure 2.18. Typical type-I (left) and type-II (right) down-converted signal-idler cones. o (e) are ordinary (extraordinary) photon cones. ph are the pinholes through which we let the photon pairs into fibers.

When $|\mathbf{k}_1| = |\mathbf{k}_2|$, the color (frequency) of the signal photon is equal to the color of the idler photon (half of the frequency of the pump photon) (see Eq. (2.116)). We achieve this outcome for a particular angle that the wave vector of the pump photon $\mathbf{k}_0$ makes with the optical axis of the crystal. Then the signal and idler photons appear at the opposite sides of the line determined by $\mathbf{k}_0$, as shown in Fig. 2.19 (a) and (c).

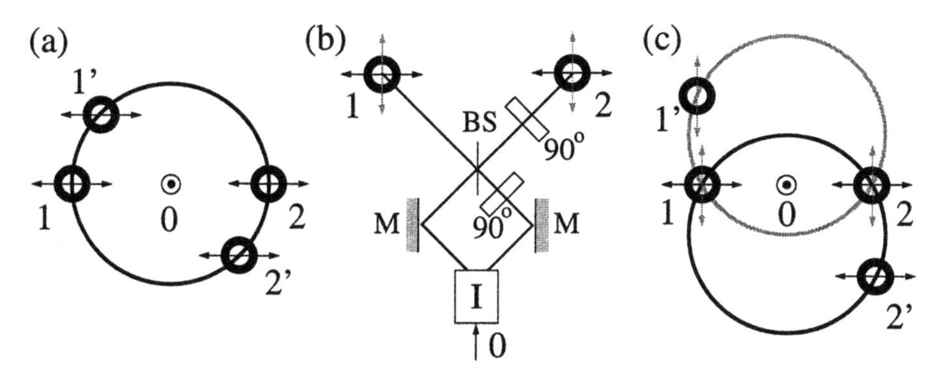

Figure 2.19. (a) Type-I down-conversion for the signal and idler of the same frequency; the cones coincide; pinholes 1–2 and 1'–2' let through the ordinary signal and idler that appear at the opposite sides of the line 0 determined by the pump wave vector $\mathbf{k}_0$; (b) EPR pair generation by means of type-I down-conversion: after the signal and idler of the same frequency interfere at the beam splitter BS, they appear entangled in a singlet-like state at pinholes 1–2; (c) EPR pair generation by means of type-II down-conversion: for the signal and idler of the same frequency the cones intersect and pinholes 1–2 let through the signal and idler entangled. Photons coming out of 1'–2' cannot entangle.

Both type-I and type-II down-converted photons can be used for the polarization-coding of entangled, EPR pairs for the BB84 protocol. With type-I down-conversion, as shown in Fig. 2.19 (b), after we rotate, say, the idler by 90°, let the signal and idler combine at a beam splitter, rotate one of the outgoing photons by 90°, and let the photons through the polarizers, according to Eq. (1.80), we get

$$\langle\Psi_t|\mathcal{D}_2^\dagger\mathcal{D}_1^\dagger\mathcal{D}_1\mathcal{D}_2|\Psi_t\rangle = \frac{1}{2}\sin^2(\theta_1-\theta_2-\frac{\pi}{2}) = \frac{1}{2}\cos^2(\theta_1-\theta_2), \qquad (2.119)$$

where Ψ_t is the triplet entangled state

$$|\Psi_t\rangle = \frac{1}{\sqrt{2}}(|0\rangle_1|0\rangle_2 + |1\rangle_1\,|2\rangle_2) \qquad (2.120)$$

that corresponds to probability (2.119).

With type-II down-conversion, as shown in Fig. 2.19(c), photons passing through the pinholes 1 and 2 also appear entangled because one cannot know which cone which photon comes from [Kwiat et al., 1995].

We see that the probability (2.119) equals the Malus law probability (1.9), so the BB84 protocol—for Alice and Bob—remains the same as presented in Sec. 1.21. However, Eve is here at a disadvantage. In Sec. 1.21 (p. 64), Alice prepares photons and sends them to Bob. As we already emphasized in Sec. 1.23 (p. 81) and in this section above, if Alice prepares photons with a polarizer and sends more than one photon per time window, these photons will all be prepared in the same way, and Eve can copy them unnoticed. If Alice and Bob use the entangled photons coming from a common source which is with Alice, then the photons Eve can catch are genuinely random, and she cannot obtain any information from them. Nevertheless, they are entangled with the photons Alice possesses and can be teleported to Bob by means of quantum repeaters (see Sec. 3.1.7, p. 151) The disadvantage of such a scheme is that the whole device is much more demanding, and proper quantum repeaters still have not been realized in a laboratory. Besides, there is a problem with both Alice and Bob catching one photon but from two different pairs within a single time window.[9] In this case the corresponding bit is wrong and they cannot find out what it is without applying the error correction scheme.

In a realistic entangled pair implementation, a phase-coding scheme will most probably again prevail (see Sec. 1.23, p. 81), since polarization is not robust enough to allow implementation over larger distances. One such scheme, according to Franson [Franson, 1989] and Tapster, Rarity, and Owens[Tapster et al., 1994], is shown in Fig. 2.20.

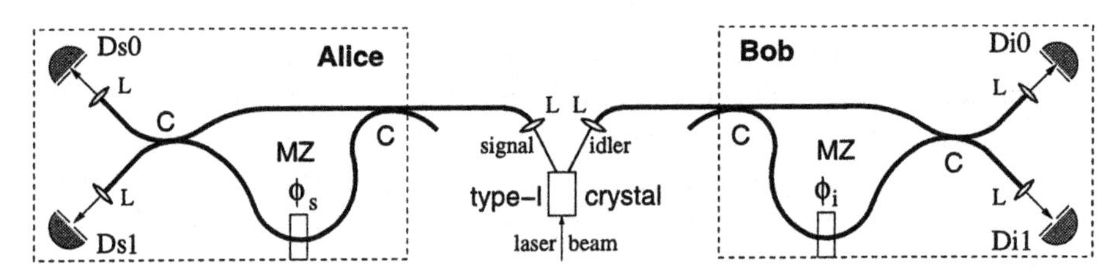

Figure 2.20. Phase-coding of entangled, EPR pair implementation of quantum key distribution according to [Tapster et al., 1994]. Fiber couplers C behave like beam-splitters and build Mach–Zehnder interferometers, MZ, on each side. Phase shifters ϕ_s and ϕ_i balance the MZs. Lenses L collect photons to the fibers and direct them towards the detectors.

The phase-coding scheme presented is based on the indistinguishability of Mach–Zehnder interferometer routes. Events where both photons pass through the long arms of the interferometer and those where they

[9] We could significantly reduce the number of such cases using two sources and double entanglement analogous to the one shown in Fig. 1.25 (p. 59) and pinholes for photons 1 and 2, say five times bigger than for photons 3 and 4 [Pavičić, 1997]. However, that would also slow down the transmission rate.

pass through the short arms cannot be distinguished from each other because the moment in which the signal and idler are down-converted is uncertain. However, when the down-conversion occurs, it occurs within femtoseconds, and we can distinguish (by precise time-of-flight measurements) the events where one photon takes the long arm and the other the short one and discard them. In a manner similar to polarization, we can calculate the coincidence rate between Ds1 and Di1 to be proportional to $\cos^2(\phi_1' - \phi_2')$, where ϕ' depends on the phase shift ϕ and the path length difference (cf. Fig. 1.32, p. 81 and Table 1.8, p. 82).

Chapter 3

PERSPECTIVES

Although Gordon Moore did not include clock speed in his law (see the Introduction), it too has increased exponentially over the years, and later on he considered it as a corollary to his law (p. xiv) so that the following formulation of his law is widely accepted in the literature today: "Computer power doubles every eighteen months" [Baxter and Trew, 2002]. In the Introduction, Sec. 1.5 (p. 9), and Sec. 1.16 (p. 36), we emphasized that by the year 2025 the shrinking of computer elements will most probably hit the quantum barrier. For the time being, the shrinking is still exponential, and the number of transistors on a single integrated-circuit chip is still doubling every 18 months. The exponential increase of the clock (processor, CPU) speed, however, stopped in mid-2002. Since then the computer power has not doubled every eighteen months any more. In Fig. 3.1, we plot the speed of the fastest CPUs of the leading world producer *Intel* against the months and years in which they were put into production according to Intel's references. From November 1971, when the first Intel 4004 processor (800 kHz) was put into production, until mid-2002, the increase in speed was exponential. After that it became linear, as shown in

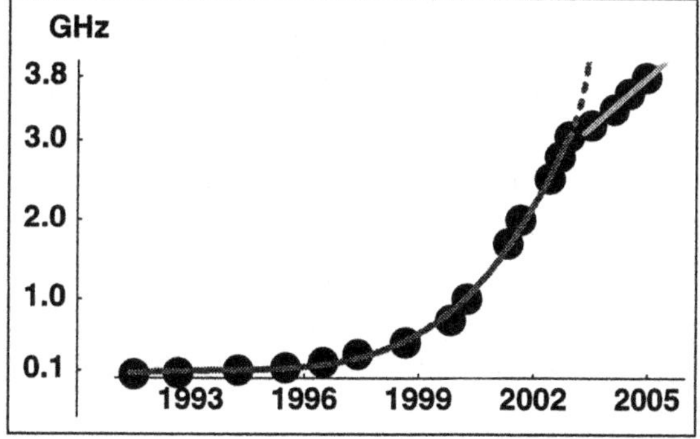

Figure 3.1: *Intel* CPU introductions: 486-50 MHz Jun 1991, DX2-66 Aug 92, P(entium)-100 Mar 94, P-133 Jun 95, P-200 Jun 96, PII-300 May 97, PII-450 Aug 98, PIII-733 Oct 99, PIII-1.0 GHz Mar 00, P4-1.7 Apr 01, P4-2.0 Aug 01, P4-2.53 May 02, P4-2.8 Aug 02, P4-3.0 Apr 03, P4-3.4 Apr 04, P4-3.6 Jun 04, P4-3.8 Nov 04.

Fig. 3.1. Moreover, "Intel [is shifting] away from clock-speed frequency as a central design philosophy... in favor of an acceleration of dual-core designs" [Krazit, 2004].

Multicore and *hyperthreading* are technologies that enable running more CPUs on one chip, and this new design philosophy leads us to *parallel computation.* Until recently, when we wanted to compute something faster, we looked for a faster CPU. In the absence of a faster CPU, the only way to run programs faster is to split them in subroutines that could run in parallel on many CPUs. And this requires rewriting the programs to make their splitting possible. Hence, in the realm of classical computing we must turn to software but at the same we should be aware that we will end up with an exponentially increasing number of CPUs.

Quantum computing, on the other hand, is inherently parallel, as we have already seen in our example of physical computing enabled by a superposition of quantum states. In Sec. 2.6 (p. 123), we have seen that it is expected that a 50-qubit quantum computer will exist by 2012. This means a superposition of $2^{50}{\approx}10^{15}$ states, each of which is composed of 50 $|1\rangle$ and $|0\rangle$ states. In the ion trap model (see Sec. 2.5, p. 109), for example, a gate operation by means of laser beams on targeted qubits also means an interaction with all 2^{50} states in parallel and changing their collective phonon modes. These qubits build a composite Hilbert space $\mathcal{H} = \mathcal{H}^2 \otimes \cdots \otimes \mathcal{H}^2$. The computational basis, i.e., the basis of this space, consists of the following 2^{50} vectors: $|00\cdots 00\rangle$, $|00\cdots 01\rangle,\ldots,$ $|11\cdots 11\rangle$.

To compute a function f of these states means to let the states evolve according to the time evolution unitary operator U (Schrödinger equation):

$$|i_1 i_2 ... i_{50}\rangle \longmapsto U|i_1 i_2 ... i_{50}\rangle = |f(i_1, ...i_{50})\rangle. \tag{3.1}$$

In a classical computer, we would carry out such a computation in a one-state-at-a-time sequence. In a quantum computer, we first put all the states on the left-hand side of Eq. (3.1) in a superposition of all 2^{50} basis states and then let them evolve together and in one step:

$$\sum_{i_1 i_2 ... i_{50}=0}^{1} \alpha_{i_1 i_2 ... i_{50}} |i_1 i_2 \ldots i_{50}\rangle \longmapsto \sum_{i_1, i_2, ..., i_{50}=0}^{1} \alpha_{i_1 i_2 ... i_{50}} |f(i_1 i_2 \ldots i_{50})\rangle.$$

After that we let the obtained (evolved) superposition collapse to a particular state that we read as a result. Of course, since such a collapse of the wave packet is intrinsically statistical, we have to repeat it a number of times but this procedure is of a polynomial complexity provided that we find a proper f for a problem we want to calculate.

In our physical computing scheme from Sec. 1.16 (p. 36), where one qubit was a quantum computer or a quantum CPU, a 50-qubit CPU would correspond to 2^{50} 1-qubit quantum CPU states working in parallel in the classical sense. Such a massive parallelism of 2^{50} $|1\rangle$–$|0\rangle$ states corresponds—ideally and theoretically—to a computational power of about one million billion transistors, and as we have mentioned in Sec. 1.7 (p. 14), not even one billion transistors have been reached so far in any classical CPU. More realistic estimates yield that 1000 qubits would be required to outpower a classical computer (see Sec. 2.6, p. 123), but that would nevertheless be a huge increase in computing power per individual qubit.

However, to be able to use this parallelism we must—as for classical parallel systems—find appropriate quantum hardware and software solutions. Quantum computing power would depend on how well we could correct errors and faults in computation, on how well we could interconnect qubits, and on how efficient the algorithms are that we would find for them. Quantum error and fault correction procedure for quantum computation is the same as the one we presented for quantum communication in Sec. 1.22 (p. 72). Quantum interconnectivity, i.e., quantum networkability, is referred to by the 6th and 7th DiVincenzo Criteria (see Sec. 2.6, p. 123) and we review some proposals for their realization in Sec. 3.1 (p. 137). In Sec. 3.2 (p. 159) we discuss some devices that may prove significant in nondestructive handling of qubits and in nonclassical measurement arrangements. As for efficient quantum algorithms, only two convincing algorithms for classical applications have been found so far: Shor's and Grover's. In Sec. 3.3 (p. 173) we shall consider the possibility of constructing a general algorithm, i.e., a general quantum algebra, which would correspond to the Boolean algebra for single classical computers. In the end, we shall consider, in Sec. 3.4 (p. 190), whether a quantum Turing machine can serve us for finding such a general quantum algorithm.

3.1 Quantum Network

The sixth and seventh DiVincenzo criteria requires that stationary and flying qubits interconvert and be faithfully transmitted between locations. The goal is the construction of a quantum bus within a quantum computer and the transfer of unknown quantum states between quantum CPUs, i.e., the construction of a quantum network. This requires either physically moving the individual qubits through a device or moving qubit states through a device. Since, according to Theorem 1.4 (p. 58), states cannot be cloned, such a transfer of states should be a teleportation of quantum states preferably "on demand."

The ability to convert qubits stored at specific points in a computer into flying qubits will be advantageous for scale-up and error correction. The question, then, is how to transfer the information stored in a fixed qubit to a flying qubit at a definite time, where the flying qubit is most often assumed to be a photon.

As we have already mentioned when considering photon-on-demand sources, the biggest problem with these sources is their poor collection efficiency (p. 128). This problem was recently resolved by means of the so-called *one-atom laser*, which contains only one photon at a time in its output beam and, like an ordinary laser, has an output within a very narrow solid angle. It has been constructed with the help of a linear ion trap device that we elaborated on in Sec. 2.5 (p. 110) and an optical cavity (optical or Fabry–Perrot resonator) which we will elaborate on in the next section). Let us, however, first take a look at some basics of the *l*ight *a*mplification of *s*timulated *e*mission of *r*adiation (*laser*).

3.1.1 Laser

As its name suggests, a laser is based on a mechanism that we use to enhance stimulated photon emission. Let us consider electrons (within an atom) that get excited (from the ground state g to level e in Fig. 3.2(a)) by means of photon absorption (that is, we will not consider other forms of energy, for example, thermodynamic, that might serve the same purpose). We say that external photons *pump* the electrons. After a few nanoseconds, the electrons make a transition to lower states. One of these lower states has to be a long-lived state that we can use for stimulated emission. These are called *metastable states* [m in Fig. 3.2 (a)]. Their average lifetime before spontaneous emission occurs is up to a millisecond—long enough to achieve stimulated emission by means of photons of frequency $\nu_{mg} = (E_m - E_g)/h$. Such a possibility was first recognized in 1924 by Richard C. Tolman as a *negative absorption* that can "reinforce the primary beam" [Tolman, 1924]—i.e., duplicate the incoming photon, as shown in Fig. 3.2 (a).

With a sufficiently intensive pump beam for exciting electrons from the ground state, we can have many more atoms in the excited metastable state than in the ground state. Such a condition is called a *population inversion*. We achieve our goal of amplifying this population with an *optical* (*resonant*) *cavity* consisting of one perfect and one partially reflecting mirror, between which we put the stimulated atoms as shown in Fig. 3.2(b). The cavity has to select out just those photons of wavelength $\lambda_{em} = c/\nu_{em}$. This is achieved by choosing the distance between the mir-

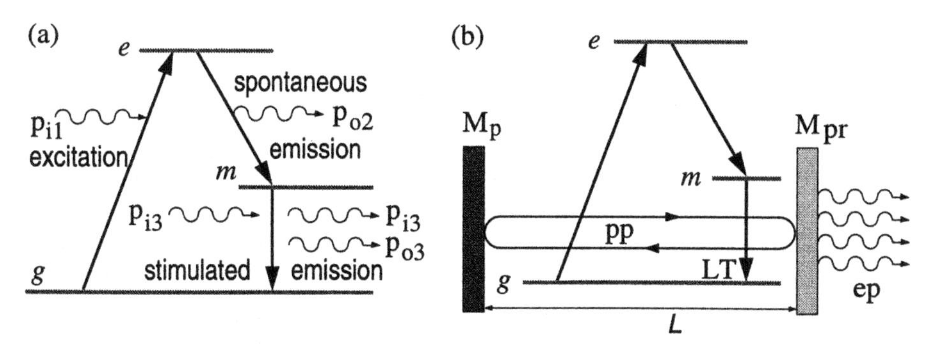

Figure 3.2. Laser basics. (a) Photon p_{i1} from the first input pump beam excites the electron from the ground state g to the excited state e. Photon p_{i3} of frequency $\nu_{mg} = (E_m - E_g)/h$ (E_m is the energy of the metastable level m) from the second input pump beam stimulates the emission of another photon p_{o3} of same frequency. (b) pp represents the photon beam traveling back and forth inside the cavity—these photons also cause the stimulated emission of new photons to join pp. i.e., to play the role of the p_{i3} beam in (a). M_p is the perfect mirror, LT denotes the laser transition, M_{pr} is a partially reflecting mirror, and ep is the exiting laser beam.

rors so as to allow standing waves only for this wavelength ($L = n\lambda_{em}/2$, where n is integer). The result is a constructive superposition of waves and a narrowly focused photon beam, because photons that are not strictly perpendicular to the mirror will be reflected out from the cavity. This situation also increases the stimulated emission until an *equilibrium state* is reached. Then the partially reflecting mirror gives a collimated *coherent* (in phase) output beam with a constant intensity (reflecting the rest back into cavity). In other words, we obtain the desired amplified stimulated emission of photons, i.e., the laser.

3.1.2 One-Atom Laser and Atom-Cavity Coupling

We will now try to describe how we can construct a one-atom laser. To make its blueprint, we have to see what we can keep and what we must change from the many-atom laser design described above and how we can justify the name "laser" for the blueprint. First, we cannot have stimulated emission any more, because we now want to get only one photon in the outgoing beam. Then, we have to revise our picture of photons moving back and forth in a cavity. In the absence of a semiclassical picture for the behavior of an atom in a cavity, we just "calculate its behavior" [Mu and Savage, 1992; Löfler et al., 1997].

The calculation makes use of the Jaynes–Cummings Hamiltonian (cf. p. 116), which will take us—as in Sec. 2.5 (p. 109)—to its ion trap implementation below. The main observable we seek to obtain is the *net stimulated emission rate*—the difference between the stimulated emission and absorption rate. The calculation yields that under realistic conditions that can be achieved in a laboratory, the net stimulated emission rate can be more than ten times greater than the spontaneous emission. This finding, together with the definite direction of the emitted photon, shows that we can give the name "laser" to a stimulated atom in a cavity. Note that we cannot use any semiclassical picture here, because there is, for example, no quantum object (for instance, another photon) that could stimulate the emission of the photon in the cavity. The sheer presence of the cavity determines the behavior of the atom and its emission of a photon at a particular rate. We call this interaction an *atom–cavity coupling.*

3.1.3 Single Photons on Demand

The next step towards realization of a device that could transfer quantum information from a stationary to a flying qubit is to keep the stationary qubit at a well defined position. We can do this with a charged atom, i.e., an ion, in an ion trap such as the one we described in Sec. 2.5. (A neutral atom is much harder to control.) Metastable to ground level transitions are no longer important because our aim is no longer an *inverted population* (see Sec. 3.1.1, p. 138) for a number of electrons. There is only one electron. Thus, a recent experiment [Keller et al., 2004a] starts with the three states of a calcium, ion as shown in Fig. 3.3.

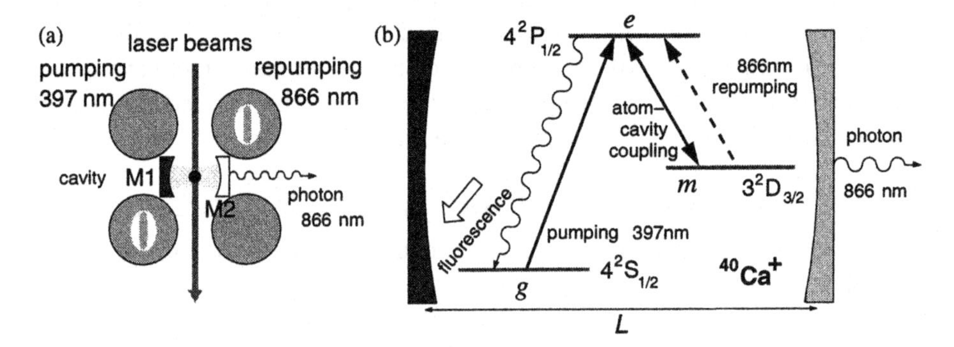

Figure 3.3. (a) An ion trapped in a linear Paul trap (see Fig. 2.14, p. 109) and coupled to a cavity as a source of single photons "on demand." M1 is a perfect and M2 a partially reflecting mirror. (b) The transition scheme of the ion in the cavity; g, e, and m are ground, excited, and metastable states, respectively.

A single calcium ion is confined in a linear trap with a lifetime on the order of hours. A spherical-mirror cavity is used because a planar-mirror one would be too sensitive to misalignment. A laser beam at a wavelength of 397 nm (ultraviolet) excites the electron from the level $4^2\text{S}_{1/2}$ to the level $4^2\text{P}_{1/2}$ in a process (Rabi oscillation) described on p. 113. The duration of the laser pulse is up to 6 μs. So, the electron can only make a transition to the metastable state $3^2\text{D}_{3/2}$ (which is not under the Rabi regime at that moment). This yields an 866 nm (infrared) photon in a cavity mode—the ion is coupled to the cavity at the resonance frequency, and the emission of the photon (decaying from the cavity) is enhanced.[1] Then another laser pulse of duration 0.5 μs pumps the electron back to $4^2\text{P}_{1/2}$. The photons from the beam are not coupled to the cavity, so they cannot contribute to the cavity's output. The electron makes a transition $4^2\text{P}_{1/2} \rightarrow 4^2\text{S}_{1/2}$ by emitting a fluorescence photon, and the sequence starts again. The sequence is repeated at a rate of 100 kHz. In other words, a continuous generation of single photons within a controlled time window has been engineered—in the Max Planck Institute in Garching [Keller et al., 2004b]. Another such deterministic single-photon generation has been independently carried out at Caltech [McKeever et al., 2004]. The setup for the latter experiment, shown in Fig. 3.4, indicates its current complexity.

Figure 3.4. Caltech Quantum Optics experimental setup of a single atom coupled to a cavity [Buck, 2003].

Such a device can be used for transferring the state of one qubit to another, although we have to adjust it for that purpose. The adjustment consists in reducing to a minimum the time the electron spends in the excited state so as to enable a full and continuous superposition between the two stable states. The goal can be achieved by the so-called adiabatic transition.

[1] Note that a semiclassical picture cannot be applied here either. We cannot say that there is a photon in a cavity leaving the ion and traveling back and forth at the speed of light in the cavity. The photon becomes "real" only when it leaves the cavity.

3.1.4 Laser Dark States

Let us modify our system so as to make it use two ground states, say Zeeman states, instead of a ground state and a metastable state. In this way we make the corresponding energies close to each other. The Schrödinger equation of our ion, with the laser beams applied to it and before it is put in the cavity, is completely analogous to Eq. (2.63), only we now have three states and will therefore obtain three equations instead of the two given by Eq. (2.73). The three-level ion system is given in Fig. 3.5 (a).

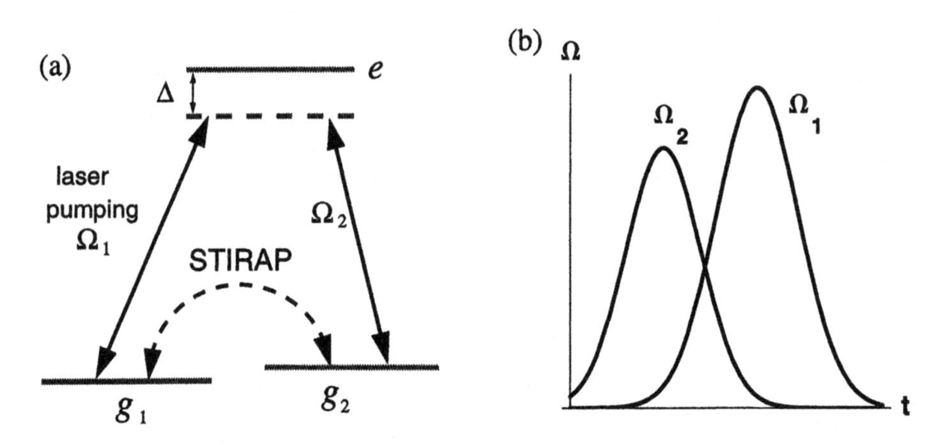

Figure 3.5. (a) A three-level ion system— the two lower levels are Zeeman splittings of the ground state of the ion. Two laser beams (Ω_1, Ω_2), which can induce transitions $g_1 \to e$ and $g_2 \to e$, respectively, interact with the system. In the STIRAP regime (see p. 144), the electron goes directly from g_1 to g_2. (b) A sequence of laser beams that induces a STIRAP transition $g_1 \to g_2$.

The matrix form of the equations is given below [Hioe, 1984; Kuklinski et al., 1989]. We obtain it in the manner described on p. 114. The only difference is that this three-level system is now driven by delayed laser pulses that we switch on and off. Therefore, the electric vectors $\mathbf{E}_1$ and $\mathbf{E}_2$ of the corresponding laser beams must be time dependent. This outcome (see p. 114) makes $\Omega_1(t)$ and $\Omega_2(t)$ time dependent as well.

Our Schrödinger equation now reads (cf. Eq. (2.63))

$$H\Psi = i\hbar\frac{\partial\Psi}{\partial t} \tag{3.2}$$

and its general solution is [cf. Eq. (2.64)]

$$|\Psi\rangle = c_{g_1}(t)e^{-i\omega_{g_1}t}|g_1\rangle + c_e(t)e^{-i\omega_e t}|e\rangle + c_{g_2}(t)e^{-i\omega_{g_2}t}|g_2\rangle. \tag{3.3}$$

Following a procedure analogous to the one used on pp. 114–116, we get the following three equations (cf. Eq. (2.73)):

$$\begin{bmatrix} \frac{dc_{g_1}}{dt} \\ \frac{dc_e}{dt} \\ \frac{dc_{g_2}}{dt} \end{bmatrix} = \frac{i}{2} \begin{bmatrix} 0 & \Omega_1(t)e^{-i\Delta t} & 0 \\ \Omega_1(t)e^{i\Delta t} & 2\Delta e^{i\Delta t} & \Omega_2(t)e^{i\Delta t} \\ 0 & \Omega_2(t)e^{-i\Delta t} & 0 \end{bmatrix} \begin{bmatrix} c_{g_1} \\ c_e \\ c_{g_2} \end{bmatrix}, \tag{3.4}$$

where $\Delta = (\omega_e - \omega_{g_1}) - \omega_1 = (\omega_e - \omega_{g_2}) - \omega_2$; and ω_1 is the frequency of the first laser beam and ω_2 of the second one—traditionally named the *Stokes laser beam.* The 3×3 matrix is equal to $-2iH/\hbar$ for $\Delta t = 0$.

From Eq. (3.4) it follows that the evolution of both c_{g_1} and c_{g_2} (and therefore of $|g_1\rangle$ and $|g_2\rangle$) depend on c_e (i.e., on $|e\rangle$). However, as we mentioned above, we are primarily interested in an evolution of $|g_1\rangle$ into $|g_2\rangle$ from which $|e\rangle$ could be excluded, and in a superposition of $|g_1\rangle$ into $|g_2\rangle$ on which $|e\rangle$ would have as little influence as possible. So we must find an eigenfunction (p. 26) of H that depends only on $|g_1\rangle$ into $|g_2\rangle$. Actually, one can verify that our Hamiltonian in the rotating wave approximation (p. 115),

$$H = -\frac{\hbar}{2} \begin{bmatrix} 0 & \Omega_1(t) & 0 \\ \Omega_1(t) & 2\Delta & \Omega_2 \\ 0 & \Omega_2(t) & 0 \end{bmatrix}, \tag{3.5}$$

has the following three eigenstates [Kuklinski et al., 1989; Bergmann et al., 1998]:

$$\begin{aligned} |\Psi^+\rangle &= \sin\Theta\sin\Phi|g_1\rangle + \cos\Phi|e\rangle + \cos\Theta\sin\Phi|g_2\rangle \\ |\Psi^0\rangle &= \cos\Theta|g_1\rangle - \sin\Theta|g_2\rangle \\ |\Psi^-\rangle &= \sin\Theta\cos\Phi|g_1\rangle - \sin\Phi|e\rangle + \cos\Theta\cos\Phi|g_2\rangle, \end{aligned} \tag{3.6}$$

where the time-varying angles Θ (usually called the *mixing angle*) and Φ are defined by means of the following equations:

$$\tan\Theta = \frac{\Omega_1(t)}{\Omega_2(t)}, \qquad \tan\Phi = \frac{\sqrt{\Omega_1^2(t) + \Omega_2^2(t)}}{\sqrt{\Omega_1^2(t) + \Omega_2^2(t) + \Delta^2} - \Delta}. \tag{3.7}$$

We see that the eigenvector $|\Psi^0\rangle$ is independent of Δ and that the corresponding eigenvalue is zero and therefore independent of Ω. In other words, $|\Psi^0\rangle$ does not depend on level e, and we therefore call it a *dark state* because it is not affected by a spontaneous emission (fluorescence) from the level e. Let us analyze the dynamics of this state. We assume that at the beginning, the electron is at level g_1. We can describe its population transfer from level g_1 to level g_2 as

$$\left.\frac{\Omega_1(t)}{\Omega_2(t)}\right|_{t\to-\infty} \to 0 \qquad \text{and} \qquad \left.\frac{\Omega_2(t)}{\Omega_1(t)}\right|_{t\to+\infty} \to 0. \tag{3.8}$$

This yields

$$\left|\langle g_1|\Psi^0\rangle\right|^2 = 1 \quad \text{for} \quad t\to-\infty, \qquad \left|\langle g_2|\Psi^0\rangle\right|^2 = 1 \quad \text{for} \quad t\to+\infty. \tag{3.9}$$

From Eq. (3.8), we see that we first have to send in an Ω_2 (Stokes) laser pulse—although the level g_2 is still unpopulated—if we want to end up with the electron populating level g_2. Also, we have to switch off the second (Stokes) laser before the first one (Ω_1).[2] This process is shown in Fig. 3.5 [Kuklinski et al., 1989; Gaubatz et al., 1990].

The next result is that the state vector we started with,

$$|\Psi(t)\rangle = |g_1\rangle \quad \text{for} \quad t\to-\infty, \tag{3.10}$$

can evolve *adiabatically*, i.e., so as to satisfy

$$|\Psi(t)\rangle \approx |\Psi^0(t)\rangle \tag{3.11}$$

at all times. This is possible under the following condition:

$$T\sqrt{\Omega_1^2+\Omega_2^2} \gg 1, \tag{3.12}$$

where T is the pulse length [Kuklinski et al., 1989; Bergmann et al., 1998]. In this way, we can have an adiabatic evolution with long pulses, strong pulses, or both. This adiabatic evolution is also called *adiabatic passage* or *stimulated Raman adiabatic passage*, STIRAP (see Fig. 3.5).

3.1.5 Cavity Dark States

To build our device for communication between two ion qubits via flying photon qubits, we put our system in a cavity in order to substitute its field for the Stokes laser field and to obtain the decayed photon in a predetermined direction. If the cavity is tuned to the same frequency as the Stokes laser field was in Sec. 3.1.4 (p. 142), it stimulates the photon emission in the same way. So we can describe the evolution of

[2]The opposite sequence would start with $\cos\Theta = 0$, that is, with $\left|\langle g_1|\Psi^0\rangle\right|^2 = 0$, which clashes with the assumed initial state in which the electron populates the g_1 level.

the trapped atom by means of the following Hamiltonian [Kuhn et al., 1999]:

$$H = -\frac{\hbar}{2}\begin{bmatrix} 0 & \Omega & 0 \\ \Omega & 2\Delta & 2g \\ 0 & 2g & 0 \end{bmatrix}, \tag{3.13}$$

in which we substitute the *atom–cavity coupling* constant g multiplied by 2 for the Stokes laser Rabi frequency Ω_2 in Eq. (3.5) and denote the Rabi frequency Ω_1 as Ω. The atom–cavity coupling constant in a semiclassical representation corresponds to the potential energy of the induced[3] atomic *electric dipole moment* $\boldsymbol{\varepsilon}$ in the single-photon cavity field [James, 1998]:

$$g = \boldsymbol{\varepsilon} \cdot \mathbf{E}_0 = \sqrt{\frac{\hbar\omega}{2\epsilon_0 V_{\text{cavity}}}}, \tag{3.14}$$

where V_{cavity} is the cavity mode volume. The closer the cavity mirrors are to each other (smaller V_{cavity}), the stronger the atom–cavity coupling.

The dark state $|\Psi^0\rangle$ remains the same as in Eq. (3.6), only instead of $\tan\Theta$ from Eq. (3.7), we now have

$$\tan\Theta = \frac{\Omega(t)}{2g}. \tag{3.15}$$

The dynamics remains the same as described in Sec. 3.1.4 (p. 142) and shown in Fig. 3.5, (p. 142 except that role of the Stokes laser field (Ω_2) is taken over by the cavity field. Thus [see Fig. 3.5 (b), p. 142], the cavity field is always "on" before the first laser beam ($\Omega = \Omega_1$) illuminates the atom. Since the excitation of the electron ($g_1 \rightarrow e$) is avoided by STIRAP we also do not have an energy loss by spontaneous emissions ($e \rightarrow g_1$ or $e \rightarrow g_2$). This also means that the STIRAP transition ($g_1 \rightarrow g_2$) is a unitary process.

A semiclassical interpretation of the role of Δ-detuning runs as follows. The first laser field excites the electron up to the $e - \Delta$ level, which is not a quantum level, from which the electron could deexcite through a spontaneous process with a photon emission to any level. However, in the presence of the cavity field that stimulates the transition $(e-\Delta) \rightarrow g_2$, the electron is forced to g_2. During this second part of the STIRAP transition, the electron generates a photon inside the cavity that subsequently decays from the cavity through the output coupler (partially reflecting mirror).

[3]Induced by the single-photon cavity field.

We will denote the initial, coupled, and final electron-photon states of the STIRAP process as

$$|g_1, 0\rangle, \qquad |g_2, 1\rangle, \qquad \text{and} \qquad |g_2, 0\rangle, \tag{3.16}$$

respectively. Here, $|g_1, 0\rangle$ means that the electron is at level g_1 and no photon is present, $|g_2, 1\rangle$ means the electron is in a transition to g_2 and a photon is generated inside the cavity, while $|g_2, 0\rangle$ means the electron is at level g_2 and the photon has decayed (was emitted) from the cavity. The dark state can be written as [see Eq. (3.6) and (3.15)]

$$|\Psi^0\rangle = \frac{1}{\sqrt{4g^2 + \Omega(t)}}(2g|g_1, 0\rangle - \Omega(t)|g_2, 1\rangle), \tag{3.17}$$

where we used the following elementary trigonometric relations:

$$\cos\Theta = \frac{1}{\sqrt{1 + \tan^2\Theta}}, \qquad \sin\Theta = \frac{\tan\Theta}{\sqrt{1 + \tan^2\Theta}}. \tag{3.18}$$

STIRAP assumes a slow rising of the first pump pulse, and the emission $|g_2, 1\rangle \rightarrow |g_2, 0\rangle$ starts as soon as the decaying state $|g_1\rangle$ starts to contribute to $|\Psi^0\rangle$, i.e., with the rising edge of the pump pulse (Ω_1 edge in Fig. 3.5 (b), p. 142) [Kuhn et al., 2002]. After the emission $|g_2, 1\rangle \rightarrow |g_2, 0\rangle$ finishes, we can re-pump the electron back to $|g_1, 0\rangle$ and start the sequence again. With a suitable choice of laser pulses, cavity, and atom, we can achieve a photon decay probability distribution of about 2 μs and repeat the sequence every 4 μs. This enables us to achieve a controlled emission of photons and a communication of two atom–cavity systems as shown in Sec. 3.1.6.

3.1.6 Dark-State Teleportation

The strongly coupled atom–cavity system photon emission based on the STIRAP unitary process, which we described in the previous section, can be reversed. Thus a superposition of ground states of one atom in one cavity can be adiabatically transfered to another such atom in another cavity by means of a photon emitted from one cavity and sent to another [Pellizzari, 1997; Pellizzari, 1998]. Such a photon communication between atoms establishes a mode of quantum networking.

Another possible building of a quantum network among atoms in cavities is by means of joint detection of photons decaying from distinct cavities and teleportation of the state of one atom to another [Yu et al., 2004]. This approach has the advantage that we do not have to carry out a demanding feeding of one cavity with a photon that has decayed from another.

Our aim is to transfer an unknown state, a superposition state of an electron in atom A in cavity A:

$$\alpha|g_1\rangle_A + \alpha|g_2\rangle_A, \tag{3.19}$$

to a state of an electron in atom B in cavity B:

$$\alpha|g_1\rangle_B + \alpha|g_2\rangle_B. \tag{3.20}$$

Therefore, $|g_1\rangle$ and $|g_2\rangle$ cannot themselves be dark states as in Eq. (3.17). Instead, we have to introduce new states that will form dark states with them. The teleportation we will carry out by means of such dark states we call *dark-state teleportation.* To arrive at the desired dark states, we pick up a suitable atom that has enough states for optical pumping. These atoms are usually alkali-metal atoms, say rubidium or cesium, which are widely used in various applications of optical pumping (atomic clocks, weak magnetic field measurements, NMR, MRI, etc.).

Following the given proposal [Yu et al., 2004], we will consider the rubidium isotope ^{87}Rb. It has the closed shells $nl = 1s, 2s, 2p, 3s, 3p, 3d, 4s$, $4p$ and one electron in the $5s$ shell, which is pushed below the $4d$ and $4f$ shells by the spin–orbit interaction (see p. 91). Thus ^{87}Rb behaves like a system with one electron in the $5s$ ground state. The total angular momentum is given by Eq. (2.11): $\mathbf{J}=\mathbf{L}+\mathbf{S}$. For the ground state $5s$, we have $s = 1/2$ and $l = 0$ and therefore $j = 1/2$. The first excited states are the $5p$ states $5p_{1/2}$ and $5p_{3/2}$, corresponding to $s = 1/2$, $l = 1$, $j = 3/2$, and $j = 5/2$, respectively. They are separated by the spin–orbit interaction $\mathbf{L} \cdot \mathbf{S}$. We will consider only $j = 3/2$.

The total nuclear angular momentum $\mathbf{K}$ combines with $\mathbf{J}$ to give the total angular momentum of the atom: $\mathbf{F} = \mathbf{J} + \mathbf{K}$. ^{87}Rb has $K = 3/2$, and its $j = 1/2$ ground states are split by the hyperfine interaction (see Eq. (2.31)) into doublets with $F = K \pm j = 3/2 \pm 1/2 = 2, 1$. Now we apply an external magnetic field $\mathbf{B}$ to the atom to split the levels into magnetic Zeeman sublevels (see Sec. 2.2, p. 88) with magnetic quantum numbers $m = -F, -F+1, \ldots, F$. The levels are given in Fig. 3.6 (cf. [Sarkisyan et al., 2005]).

As we have shown in Sec. 1.9 (p. 20), a circularly polarized photon has an angular momentum $j_p = 1$ and two additional degrees of freedom (eigenvalues of $\mathbf{k} \cdot \mathbf{j}_p/k$) denoted $m_{j_p} = \pm 1$. The linear polarization of photons that we have used so far is a superposition of their circular polarizations as given by Eq. 1.13. We cannot use linearly polarized photons here because the selection rules for them require $\Delta m = 0$ instead of $\Delta m = \pm 1$ in Eq. (3.21).

When our atom absorbs a circularly polarized photon, it has to absorb both energy and angular momentum in its transition from the ground

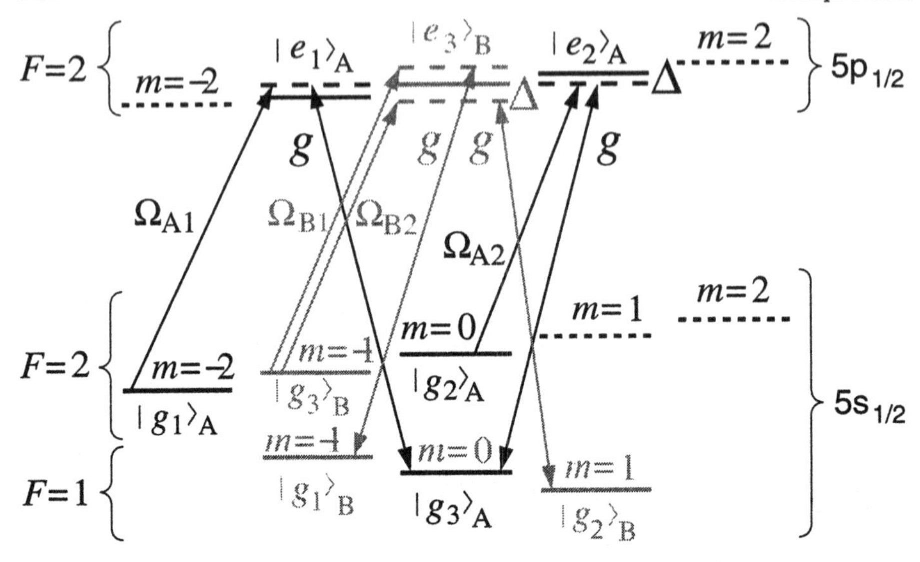

Figure 3.6. Levels of ^{87}Rb ($|g\rangle$, $|e\rangle$), laser beams (Ω), atom–cavity couplings (G), and detunings (Δ) relevant to the teleportation experiment. Atoms A and B use different levels, so we show them together here (A—black, B—gray). Dotted levels are not used. Dashed levels denote detuning.

to an excited state, and therefore the following selection rules must be satisfied:

$$\Delta l = \pm 1, \qquad \Delta F = \pm 1, \qquad \Delta m = m_{j_p} = \pm 1. \tag{3.21}$$

By emission of a photon, the same selection rules must be satisfied. Thus, for $\Delta m = \pm 1$ we get a circularly polarized photon and for $\Delta m = 0$ a linearly polarized photon.

In the same way as in Secs. 3.1.4 (p. 142) and 3.1.5 (p. 144), we make transitions $|g_1\rangle_A \to |e_1\rangle_A$, ($\Omega_{A1}$), $|g_2\rangle_A \to |e_2\rangle_A$, ($\Omega_{A2}$), and $|g_3\rangle_B \to |e_3\rangle_b$ (Ω_{B1} and Ω_{B2} because of their opposite detunings) by means of right-hand circularly polarized laser beams (note that all transitions have $\Delta m = +1$ and see Eqs. (3.21) and (1.12)).

The transitions $|e_1\rangle_A \to |g_3\rangle_A$ and $|e_3\rangle_B \to |g_2\rangle_B$ are coupled to cavity mode a_L with a left-hand circular polarization, while $|e_2\rangle_A \to |g_3\rangle_A$ and $|e_3\rangle_B \to |g_1\rangle_B$ are coupled to a_R with a right-hand one. The atom-cavity constants for the first and second cavities are equal so as to give photons of the same frequency and enable their subsequent interference. In analogy with Eq. (3.17), we express the state of system A described by the two dark states (recall Eq. (3.19)):

$$\begin{aligned}|\Psi(t)\rangle_A &= \frac{a}{\sqrt{4g^2+\Omega_{A1}(t)}}(2g|g_1,0\rangle_A - \Omega_{A1}(t)|g_3,L\rangle_A)\\ &+ \frac{b}{\sqrt{4g^2+\Omega_{A2}(t)}}(2g|g_2,0\rangle_A - \Omega_{A2}(t)|g_3,R\rangle_A). \qquad (3.22)\end{aligned}$$

Within the STIRAP process, with Ω_{A1} and Ω_{A2} increasing gradually, $|\Psi(t)\rangle_A$ adiabatically evolves into

$$|\Psi\rangle_A = a|g_3,L\rangle_A + b|g_3,R\rangle_A). \qquad (3.23)$$

The state of system B is

$$\begin{aligned}|\Psi(t)\rangle_B &= \frac{1}{\sqrt{4g^2+\Omega_{B1}(t)}}(2g|g_3,0\rangle_B - \Omega_{B1}(t)|g_1,R\rangle_B)\\ &+ \frac{1}{\sqrt{4g^2+\Omega_{B2}(t)}}(2g|g_3,0\rangle_B - \Omega_{B2}(t)|g_2,L\rangle_B), \qquad (3.24)\end{aligned}$$

and it too, with Ω_{B1} and Ω_{B2} increasing gradually, adiabatically evolves into

$$|\Psi\rangle_B = \frac{1}{\sqrt{2}}(|g_1,R\rangle_B + |g_2,L\rangle_B). \qquad (3.25)$$

The joint state of systems A and B is

$$\begin{aligned}|\Psi\rangle_{AB} = |\Psi\rangle_A \otimes |\Psi\rangle_B &= a\,|g_3,L\rangle_A|g_1,R\rangle_B + a|g_3,L\rangle_A|g_2,L\rangle_B\\ &+ b|g_3,R\rangle_A|g_1,R\rangle_B + b|g_3,R\rangle_A|g_2,L\rangle_B. \qquad (3.26)\end{aligned}$$

Our goal is to detect just two of the four photons and to infer the state of atom B from the measurement. More specifically, we want one photon decaying from cavity A and one from cavity B to combine at a polarized beam splitter, as shown in Fig. 3.7. [This is why g in Eq. (3.22) is equal to g in Eq. (3.24).] Since we do not know which of the two photons from each cavity would decay first, we have to represent each photon as a superposition of two possible photon states. So the joint state of the two photons leaving the cavities is [cf. Eq. (1.68)]

$$\begin{aligned}|\psi\rangle_{AB} &= (a|L\rangle_A + b|R\rangle_A) \otimes (|L\rangle_B + |R\rangle_B)\\ &= a|L\rangle_A|L\rangle_B + a|L\rangle_A|R\rangle_B + b|R\rangle_A|L\rangle_B + b|R\rangle_A|R\rangle_B. \qquad (3.27)\end{aligned}$$

To carry out the measurement, it is convenient to turn the circular polarization into the linear one. Quarter-wave plates, QWPs, with additional rotators do just this: $L \to H$, $R \to V$, where H and V mean

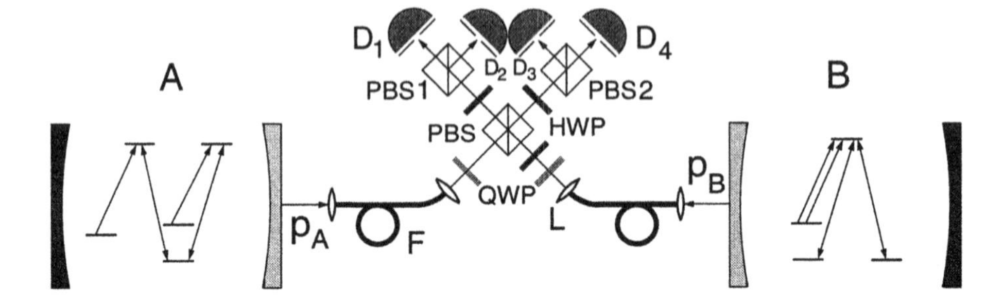

Figure 3.7. Teleportation of the state of atom A to the state of atom B via photons decaying from cavities A and B. PBS, polarizing beam splitters; L, lenses; F, fiber loops; D, detectors; and p, photons decaying from the cavities. QWP and HWP are quarter- and half-wave plates (p. 22). The QWPs turn circular polarizations into linear ones. The HWP between PBS and QWP is a 90° plate, and the HWPs between PBS and PBS1, PBS2 are 45° plates.

horizontal and vertical polarization (see p. 22). The right half-wave plate (in front of the polarizing beam splitter PBS) then rotates the polarization plane by 90° (see p. 22). This transforms Eq. (3.27) into

$$|\psi\rangle_{AB} = a|H\rangle_A|V\rangle_B + a|H\rangle_A|H\rangle_B + b|V\rangle_A|V\rangle_B + b|V\rangle_A|H\rangle_B. \quad (3.28)$$

A polarizing beam splitter transmits only the horizontal polarization component and reflects the vertical one. Thus we have $|H\rangle_A$ going towards PBS2, $|V\rangle_A$ towards PBS1, $|H\rangle_B$ towards PBS1, and $|V\rangle_B$ towards PBS2. We shall denote them $|H_2\rangle$, $|V_1\rangle$, $|H_1\rangle$, and $|V_2\rangle$, respectively. Hence, immediately after PBS in Fig. 3.7, $|\psi\rangle_{AB}$ evolves into [Pan and Zeilinger, 1998]

$$a|H_2\rangle|V_2\rangle + a|H_1\rangle|H_2\rangle + b|V_1\rangle|V_2\rangle + b|V_1\rangle|H_1\rangle. \quad (3.29)$$

Let us consider only the case where one photon goes through PBS1 and the other through PBS2:

$$|\psi\rangle_{12} = a|H_1\rangle|H_2\rangle + b|V_1\rangle|V_2\rangle = \frac{1}{\sqrt{2}}[(a+b)|\phi^+\rangle + (a-b)|\phi^-\rangle], \quad (3.30)$$

where

$$|\phi^+\rangle = \frac{1}{\sqrt{2}}(|H_1\rangle|H_2\rangle + |V_1\rangle|V_2\rangle), \quad |\phi^-\rangle = \frac{1}{\sqrt{2}}(|H_1\rangle|H_2\rangle - |V_1\rangle|V_2\rangle). \quad (3.31)$$

We can tell $|\phi^+\rangle$ from $|\phi^-\rangle$ as follows. We first rotate their polarization plane by 45° with the help of two HWPs in front of PBS1 and PBS2:

$$|H_i\rangle \to \frac{1}{\sqrt{2}}(|H_i\rangle + |V_i\rangle), \qquad |V_i\rangle \to \frac{1}{\sqrt{2}}(|H_i\rangle - |V_i\rangle), \quad (3.32)$$

where $i = 1, 2$. This causes the evolution

$$|\phi^+\rangle \rightarrow \frac{1}{\sqrt{2}}(|H_1\rangle|H_2\rangle + |V_1\rangle|V_2\rangle), \ |\phi^+\rangle \rightarrow \frac{1}{\sqrt{2}}(|H_1\rangle|V_2\rangle + |V_1\rangle|H_2\rangle). \tag{3.33}$$

If the photons are in the first of these two states, they will trigger (ideally) either D_1 and D_4 or D_2 and D_3[4]—and this means that $|\psi\rangle_{12}$ in Eq. (3.30) was in the state $|\phi^+\rangle$, and therefore

$$|\Psi\rangle_{AB} = \frac{a+b}{2}(|H\rangle_A|H\rangle_B + |V\rangle_A|V\rangle_B). \tag{3.34}$$

Taking into account the rotation of the HWP in front of PBS, we see that Eq. (3.34) corresponds to a superposition of $|L\rangle_A|R\rangle_B$ and $|R\rangle_A|L\rangle_B$, which means that upon emission of these photons from the cavities, the joint state of systems A and B given by Eq. (3.26) jumps into

$$|\Psi\rangle_{AB\text{fin}} = |g_3, 0\rangle_A(a|g_1, 0\rangle_B + b|g_2, 0\rangle_B). \tag{3.35}$$

In other words, atom B was left in the state given by Eq. (3.20), and the desired teleportation has been carried out. The classical part of the teleportation procedure consists of discarding all the other states of the systems A and B for which we obtained some other clicks in detectors D_1–D_4.[5]

3.1.7 Quantum Repeaters

When teleportation devices such as the one shown in Fig. 3.7 (p. 150) or the one presented in Fig. 1.32 (p. 81) are used for communicating unknown states from one point in space to another, such devices are called *quantum repeaters.* As we already emphasized (p. 75), according to Theorem 1.4 (p. 58) we cannot copy a quantum state and therefore we cannot amplify a quantum signal, but we can teleport the signal through fibers. The question is whether we can cover a greater distance with the same losses using repeaters than with a single fiber.

It would be best if we could use entangled photons on demand, because our final aim is to integrate quantum communication and quantum computation. As we have seen in Sec. 3.1.3 (p. 140), single photons on demand have recently been obtained on devices that can be incorporated

[4]For any other set of detector clicks, we simply abort the procedure, destroy the state of system A, and start everything anew. This means that the procedure is essentially probabilistic in the same sense in which the teleportation presented in Sec. 1.20 (p. 56) is probabilistic.

[5]We can also keep the state that corresponds to the triggering of either D_1 and D_3 or D_2 and D_4, thus selecting $|\phi^-\rangle$ in Eq. (3.30). We then find the atom B in the state $a|g_1\rangle_B - b|g_2\rangle_B$, which we can bring to the state (3.20) by giving $|g_2\rangle_B$ a phase shift $[\exp(i\pi) = -1]$.

into future quantum computers. However, so far, progress in obtaining entangled photons on demand has been made almost exclusively with *quantum dots* in a cavity [Benson et al., 2000; Stace et al., 2003; Beny-oucef et al., 2004; Kumar et al., 2004]. Therefore we shall assume our sources are quantum dots, but this assumption is not essential for our elaboration.[6] It is also expected that there will be other such sources soon. On the other hand, there are several reasons why down-converted photons do not seem suitable for the purpose. First, for $n-1$ repeaters we should have n sources (say type-II crystals, p. 130), all simultaneously pumped by a laser beam. Second, we have an unsurmountable problem with the probabilistic nature of the down-conversion of signal–idler pairs—we have to use weak pump pulses to avoid down-conversion of more than one pair in a crystal and to prevent eavesdropping (p. 127), and this leads to an exponential decrease in the probability of a simultaneous down-conversion of n pairs in n crystals (this probability is equal to the probability of down-converting n pairs in one of the crystals and none in all the others).

From Sec. 1.20 (p. 56), it is clear that one can use the setup from Fig. 1.25 (p. 59) as a repeater. We just take out the polarizers $P3$ and $P4$ and use either type-I or type-II crystals as shown in Fig. 2.19 (p. 131) as our sources S_I and S_{II}. Then, as soon as $D3$ and $D4$ each detects a photon, the other two photons leaving S_I and S_{II} to the left are entangled in polarization that Alice and Bob can verify using polarizers $P1$ and $P2$ and detectors $D1$ and $D2$. This result follows from Eqs. (1.128), (1.145), (1.141), and (1.142). We have seen in Sec. 2.7 (p. pagerefsec:quant-comm-impl) that the losses in fibers rise exponentially with their length. Let us assume that we cannot have a faithful transmission for fibers longer than $L_{\max}$. If we now substitute fibers of length $L_{\max}$ for the paths from $D1$ to S_I, from S_I to BS, from BS to S_{II}, and from S_{II} to $D2$ in our repeater setup in Fig. 1.25 (p. 59) and assume that both detectors and sources are ideal and perfect, we see that this setup might enable communication over a $4L_{\max}$ distance. Thus, we are tempted to consider a concatenation of such repeaters—as shown in Fig. 3.8—as a solution to our quantum communication problem. However, the setup cannot work this way. Below we show why this is not possible and how we can construct proper quantum repeaters.

---

[6]Also, since according to Quantum Computation Roadmap (Table 2.1, p. 124) quantum dots are not considered to be among the main candidates for a future quantum computer, we will not enter into details of their physics.

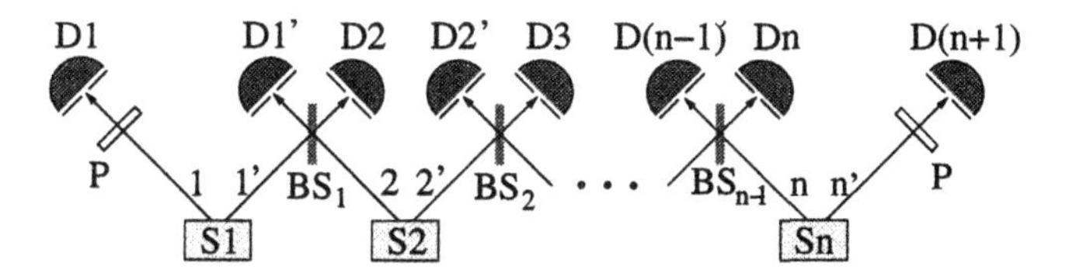

Figure 3.8. A quantum repeater scheme whose probability of a faithful transmission from 1 to n' exponentially decreases with the number of sources n [Pavičić, 2000a]. S, photon sources; P, polarizers; BS, beam splitters; and D, detectors. Photon 1 and photon n' are in a singlet state.

The sources S_1,S_2,...,S_n from Fig. 3.8 produce *nonmaximal singlets*

$$|\Psi_i\rangle = \frac{1}{\sqrt{\alpha_{ii'}^2 + \beta_{ii'}^2}} \left(\alpha_{ii'}|0\rangle_i|1\rangle_{i'} \ - \ \beta_{ii'}|1\rangle_i|0\rangle_{i'}\right), \tag{3.36}$$

where $i = 1, \ldots, n$; $|0\rangle, |1\rangle$ mean horizontal and vertical polarization respectively; and αs and βs take care of possible unknown imperfections in quantum channels and sources.

The wave function of two photon pairs born in S_1 and S_2 (Eq. (3.36)) is given by Eq. (1.145):

$$\begin{aligned} |\Psi_{12}\rangle &= |\Psi_1\rangle \otimes |\Psi_2\rangle \\ &= \alpha\,|0\rangle_1|1\rangle_{1'}|0\rangle_2|1\rangle_{2'} \ - \beta\,|0\rangle_1|1\rangle_{1'}|1\rangle_2|0\rangle_{2'} \\ &\quad -\gamma\,|1\rangle_1|0\rangle_{1'}|0\rangle_2|1\rangle_{2'} \ + \delta\,|1\rangle_1|0\rangle_{1'}|1\rangle_2|0\rangle_{2'}. \end{aligned} \tag{3.37}$$

These pairs contain photons 1' and 2, which interfere at BS_1. Upon introducing the wave function (3.37) into Eq. (1.135), we get a nonmaximal singlet state [Pavičić, 1995; Pavičić, 1997]:

$$\frac{1}{\sqrt{(\alpha_{11'}\beta_{22'})^2 + (\alpha_{22'}\beta_{11'})^2}}(\alpha_{11'}\beta_{22'}|0\rangle_1|1\rangle_{2'} - \alpha_{22'}\beta_{11'}|1\rangle_1|0\rangle_{2'}). \tag{3.38}$$

For $\alpha_{11'}\beta_{22'} = \alpha_{22'}\beta_{11'}$, we achieve a kind of *purification* [Zhao et al., 2003] of the final state (3.38) since it then becomes a maximal singlet

$$|\Psi_s\rangle = \frac{1}{\sqrt{2}}(|0\rangle_1|1\rangle_{2'} - |1\rangle_1|0\rangle_{2'}). \tag{3.39}$$

In general, we have to carry out a purification separately. It can be done by means of quantum circuits, notably by means of the so-called *bilateral* CNOT gates, BCNOT [Bennett et al., 1996]. Probabilistic all-optical CNOT gates [Pittman et al., 2001] have recently been proposed for building quantum repeaters [Kok et al., 2003]. We shall, however, present a simpler quantum optical purification [Pan et al., 2001], which is not ideally efficient but is considered for and has been experimentally realized within a quantum repeater [Zhao et al., 2003]. Its scheme is given in Fig. 3.9 (a).

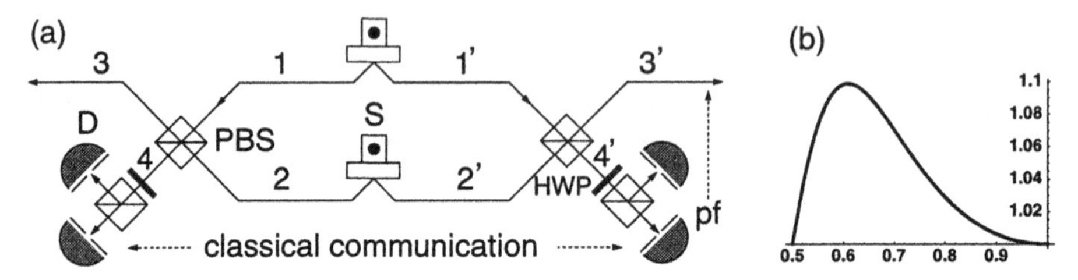

Figure 3.9. (a) A probabilistic purification scheme according to [Pan et al., 2001]. S are photon sources (quantum dots), PBS are polarizing beam splitters, D are detectors, HWP are 45° half-wave plates, and pf are phase flippers based on classical communication. Photons pairs 1-1' and 2-2' are supposed to be mostly in one of the Bell states (Eq. (3.40)) but possibly "spoiled." After a procedure that follows detections of photons 4 and 4' (see text), photons 3 and 3' will have a higher degree of entanglement than original pairs. (b) Graph of k from Eq. (3.45).

In a way, the scheme is a special case of the complete bilateral CNOT approach, which handles all *Bell states* (singlet $|\Psi^-\rangle$ and triplets $|\Psi^+\rangle$, $|\Phi^\pm\rangle$):

$$\begin{aligned}|\Psi^\pm\rangle_{ii'} &= \frac{1}{\sqrt{2}}(|0\rangle_i|1\rangle_{i'} \pm |1\rangle_i|0\rangle_{i'}),\\ |\Phi^\pm\rangle_{ii'} &= \frac{1}{\sqrt{2}}(|0\rangle_i|0\rangle_{i'} \pm |1\rangle_i|1\rangle_{i'}),\end{aligned} \tag{3.40}$$

while the scheme itself handles just two of them. It assumes that the original pairs are mostly in, say, the $|\Phi^+\rangle_{ii'}$ state with a mixture of unwanted $|\Psi^+\rangle_{ii'}$ state:

$$|\Psi_{\rm orig}\rangle_i = \frac{1}{\sqrt{F^2+(1-F)^2}}(F|\Phi^+\rangle_{ii'} + (1-F)|\Psi^+\rangle_{ii'}), \tag{3.41}$$

where F determines an unknown portion of $|\Phi^+\rangle_{ii'}$ that should be purified and $1-F$ a portion of unwanted $|\Psi^+\rangle_{ii'}$ that we want to get rid of.

The function describing two-pair state of pairs leaving sources is a product of their functions given by Eq. (3.41), i.e., the following mixture:

$$\begin{aligned}|\Psi\rangle_{\rm in} &= \frac{1}{F^2+(1-F)^2}\Big[F^2|\Phi\rangle_{11'}|\Phi\rangle_{22'} + (1-F)^2|\Psi\rangle_{11'}|\Psi\rangle_{22'}\\ &\quad + F(1-F)(|\Phi\rangle_{11'}|\Psi\rangle_{22'} + |\Psi\rangle_{11'}|\Phi\rangle_{22'})\Big]\,.\end{aligned} \tag{3.42}$$

Obviously, the middle terms cannot contribute to all four outgoing modes (3, 3', 4, 4'), because photons have equal polarizations in $|\Phi\rangle_{11'}$ and opposite one in $|\Phi\rangle_{11'}$. So, if two photons exits through, for example, 3 and 4, the other two must both exit through either 3' or 4'.

The first term reads

$$\begin{aligned}|\Phi\rangle_{11'}|\Phi\rangle_{22'} &= \frac{1}{2}\Big(|0\rangle_1|0\rangle_{1'}|0\rangle_2|0\rangle_{2'} + |0\rangle_1|0\rangle_{1'}|1\rangle_2|1\rangle_{2'} \\ &\quad + |1\rangle_1|1\rangle_{1'}|0\rangle_2|0\rangle_{2'} + |1\rangle_1|1\rangle_{1'}|1\rangle_2|1\rangle_{2'}\Big), \qquad (3.43)\end{aligned}$$

where only the first and last term correspond to only one photon going in each of the 3, 4, 3', 4' modes. This leaves us with only 50% of the photons. We can now engineer the 3–3' state by the same procedure we used in Sec. 3.1.6 on p. 150 (see Fig. 3.7, p. 150). The "only" difference is that there the photons come from the same polarizing beam splitter and here from two different ones. If the same detectors (see Figs. 3.7 (p. 150) and 3.9 (a) (p. 154), and p. 150) click, the 3–3' state is $|\Phi^+\rangle_{33'}$, and if different ones click, then the 3–3' state is $|\Phi^-\rangle_{33'}$. In the latter case, we have to perform a local phase flip operation on one of the photons (pf in Figs. 3.9 (a)) to convert $|\Phi^-\rangle_{33'}$ into $|\Phi^+\rangle_{33'}$. We obtain $|\Psi^+\rangle_{33'}$ analogously.

To estimate the quality of purification, we have to calculate the probability for the obtained $|\Phi^+\rangle_{33'}$. The probabilities for $|\Phi\rangle_{11'}|\Phi\rangle_{22'}$ and $|\Psi\rangle_{11'}|\Psi\rangle_{22'}$ are F^4 and $(1-F)^4$, respectively. Since $|\Phi^+\rangle_{33'}$ and $|\Psi^+\rangle_{33'}$ were obtained from $|\Phi\rangle_{11'}|\Phi\rangle_{22'}$ and $|\Psi\rangle_{11'}|\Psi\rangle_{22'}$, respectively, with 50% photon loss (see the previous paragraph), their probabilities are $F^4/2$ and $(1-F)^4/2$, respectively. Thus a purified state reads

$$|\Psi_{\text{purifed}}\rangle_i = \frac{1}{\sqrt{F^4 + (1-F)^4}}(F^2|\Phi^+\rangle_{ii'} + (1-F)^2|\Psi^+\rangle_{ii'}). \qquad (3.44)$$

Let us compare the coefficient of the first (purified) term with the coefficient of the first term of the original function Eq. (3.41):

$$\frac{F^2}{\sqrt{F^4 + (1-F)^4}} = k\,\frac{F}{\sqrt{F^2 + (1-F)^2}}. \qquad (3.45)$$

When we plot k on the [0.5,1] interval, we obtain the graph shown in Fig. 3.9 (b). It has a maximum for $F = 0.608423$ and it shows that the purification rate slows down to practically zero for $F > 0.9$. Hence, for a realistic application, we have to use bilateral CNOT gates as already emphasized above.

Actually, the main reason why we cannot easily have good performance with linear optical elements (beam splitters, mirrors, polarizers, phase shifters, rotators, photodetectors, delay lines, electronic switchers, etc.) and why we always lose photons is the following no-go theorem by Vaidman and Yoran [Vaidman and Yoran, 1999] and Lütkenhaus, Calsamiglia and Suominen [Lütkenhaus et al., 1999].

THEOREM 3.1 *The distinguishing of all four Bell states cannot be achieved with linear optical elements in a two-particle (4-dimensional) polarization Hilbert space.*[7]

Proof By exhaustive checking of Bell state decompositions as carried out in [Vaidman and Yoran, 1999] and [Lütkenhaus et al., 1999]. ∎

We have seen in Sec. 1.20 (p. 56) that we can identify the singlet photon state just by detecting the photons at the opposite sides of a beam splitter after they interfered at it. For all three triplet states, both photons exit from the same side of the beam splitter. By using polarizing beam splitters as shown in Fig. 3.10 (a), we can identify the triplet state $|\Psi^+\rangle$—the photons have opposite polarizations, and therefore clicks of either D_1 and D_2 or D_3 and D_4 detect $|\Psi^+\rangle$ and distinguish it from $|\Phi^\pm\rangle$. By rotating the polarizing beam splitter by 45° as shown in Fig. 3.10 (b), we can distinguish $|\Phi^\pm\rangle$ from each other but no longer from $|\Psi^+\rangle$.

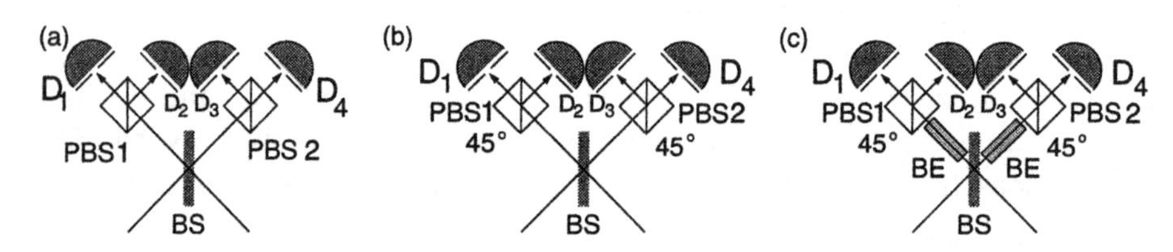

Figure 3.10. (a) A 50% *coupler* (also called a *swapper*): setup to allow identification and distinguishing of two Bell states: $|\Psi^\pm\rangle$; (b) the setup is to identify three Bell states: $|\Psi^-\rangle$ and $|\Phi^\pm\rangle$—however, it cannot distinguish $|\Phi^\pm\rangle$ from $|\Psi^+\rangle$; (c) the setup is to identify and distinguish all four Bell states in a higher Hilbert space. BE are birefringent elements.

To identify and distinguish all four Bell states, we have to embed the Bell states into a larger Hilbert space so as to obtain *hyperentangled* quantum states with more than one degree of freedom [Kwiat, 1997;

[7]There is a corresponding position-momentum problem for particle systems known as the Pauli problem: Is the knowledge of the probability distribution $|\psi(q)|^2$ and $|\psi(p)|^2$ sufficient to determine the quantum state $|\psi\rangle$ of the particle? "The question whether all real states might be Pauli unique has been answered in the negative by Pavičić [Pavičić, 1987a]... Clearly, the probability distributions $|\psi(q)|^2$ and $|\psi(p)|^2$ are not sufficient to determine the quantum state." [Weigert, 1992]

Kwiat and Weinfurter, 1998]. For example, type-II down-converted photons are polarization and momentum entangled. Such photon pairs can be passed through strongly birefringent material of length L, whose axis is set in the 0–1 linear polarization plane, with the effect of delaying one of the photons with respect to the other by the time $\Delta nL/c$, where Δn is the difference in the refractive indices. Hence, in the setup shown in Fig. 3.10 (c), we first distinguish $|\Psi^-\rangle$ from the other Bell states[8] (it is detected at the opposite sides of BS), then we temporally distinguish $|\Psi^+\rangle$ from $|\Phi^+\rangle$, and if it is $|\Phi^+\rangle$ we have to distinguish it from $|\Phi^-\rangle$ according to the procedure given on p. 150. Hyperentangled two-photon states have been realized experimentally [Cinelli et al., 2005] and mentioned in the Quantum Computation Roadmap (p. 124), but their inclusion in a quantum repeater is apparently too intricate to implement at the present time.[9] It is not only a question of sources but also of single photon detectors that can distinguish one from two photons [Kim et al., 1999; Kwiat and Weinfurter, 1998; Kok et al., 2002].

Instead, we shall present a proposal [Kok et al., 2003] that is in principle realizable with today's technology, although it is uncertain whether we can shrink it sufficiently in the near future. The proposal uses probabilistic purifiers based on all-optical realization of bilateral CNOT gates (p. 153) together with a *quantum nondemolition detection device* (QND device) [Kok et al., 2002], the probabilistic couplers presented in Fig. 3.10 (a), and a realistic quantum dot source of Bell states, as shown in Fig. 3.9. The repeater is assembled as shown in Fig. 3.11.

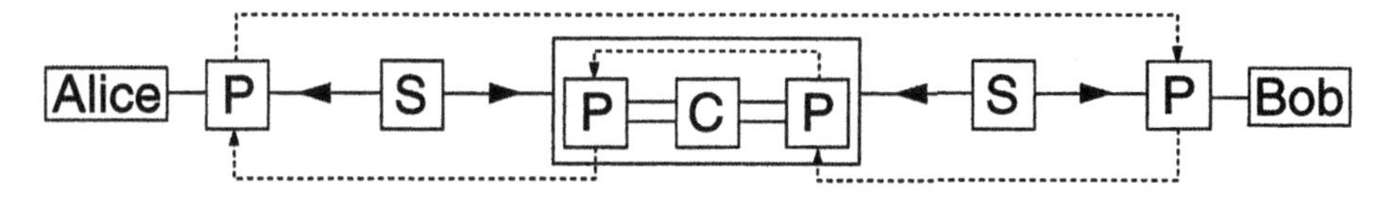

Figure 3.11. Quantum repeater according to [Kok et al., 2003]. C is a quantum coupler (swapper) as given in Fig. 3.10 (a), P in the middle block is a purifier as described in the text, and S is a quantum dot source of Bell states on demand (see p. 152 and Fig. 3.9, p. 154). Dashed lines denote classical communication.

To determine the probability of success of the repeater, we start from an essential property of the probabilistic all-optical nondestructive CNOT, namely, "that the desired CNOT operation may be performed with probability of 1/4 without measuring or determining the values of

[8]See Eq. (3.40)

[9]A similar setup for fermions (electrons) which combines spin and charge measurements has also been proposed recently [Beenakker et al., 2004].

the input qubits" [Pittman et al., 2001]. Therefore, we can concatenate as many CNOT gates within a repeater as needed to successfully perform the desired CNOT operation. Thus the total probability of success for our repeater can be obtained by multiplying the probabilities for the purifiers with the probability for the coupler.

To find the success probability for a purifier within P, we have to take into account the efficiencies of all of its elements. There are two CNOT gates with success probability p_c, one QND device with p_q, five quantum dot Bell state sources (two sources shown in Fig. 3.11, then one for each CNOT and one for the quantum nondemolition detection device—not shown) with probability p_s of successful generation of pairs, and eight detectors with efficiency η. To these we have to add a parameter ζ for the photon loss over the channel and an attenuation parameter γ. The total probability is therefore

$$p_{\text{pur}} = (1-\gamma)\zeta p_s^5 \eta^8 p_c^2 p_q \,. \tag{3.46}$$

Therefore, to obtain a probability of success for each P-box in Fig. 3.11 equal to one, we have to use $N_{\text{pur}} = 1/p_{\text{pur}}$ elements within each purifier.

The probability of success for an ideal coupler is, as we have shown above, 0.5. Taking into account the efficiency of the detectors that fire, we get $p_{\text{coupl}} = \eta^2/2$. So we have to use $N_{\text{coupl}} = 1/p_{coupl}$ couplers. Since a repeater needs two purifiers and one coupler in the central block and two purifiers at the end, the total number of components within the repeater for the values $\eta = 0.8$, $p_s = 0.9$, $\gamma = 1/2$, $\zeta = 1/2$, $p_c = 1/4$, and $p_q = 1/8$ is

$$N_{\text{total}} = 2N_{\text{pur}}(N_{\text{coupl}} + 1) = \frac{2}{p_{\text{pur}} p_{coupl}} = 10343. \tag{3.47}$$

Here we have $N_{coupl} = 4$ $(1/p_{coupl} = 3.125)$ couplers, and this implies eight sources S and a fourfold fiber. When we take into account that the number of transistors on a Pentium chip is 3.3×10^8, this number of elements might seem feasible in the near future, provided that we have single photon detectors with a realistic efficiency of 0.8 soon.

Such repeaters can easily be concatenated with polynomially increasing losses. To see this, let us look at the fidelity of photon transmission through a fiber: $F \sim \exp(-L\gamma)$, where L is the distance between Alice and Bob. To enable extraction of maximal entanglement by purifying the photon states, we must not go below a minimum fidelity F_0. Therefore, we have to install a repeater after each L/N portion of their total distance, which decreases the fidelity to F_0. This means that there are N repeaters on the total distance L, and the fidelity of each portion decreases by a factor $\alpha = \exp(-L\gamma/N)$, which represents an exponential improvement compared to the $\exp(-L\gamma)$ above. Thus each portion (each looking like the total device shown in Fig. 3.11) needs $1/\alpha$ photons, and all portions of the distance L need N/α photons, which means a polynomially increasing number of photons.

This rounds up the bits and pieces of our would-be quantum network. They still do not fit to each other smoothly, but they soon will. For example, the DARPA network (p. 125) plans to introduce quantum repeaters soon.

3.2 Quantum–Classical Coupling

In this section, we consider two couplings of quantum systems to macrodevices: interaction-free and Kochen–Specker preparations of quantum systems and their application to quantum computation.

3.2.1 Interaction-Free Computation

Quantum interference of individual systems has been found to be capable of detecting objects without transferring energy to them. "Consider a photon experiment, shown in Fig. 3.12, which results in interference in the region D provided that we do not know whether the photon arrived at the region by path s_1 or by path s_2 ... If we, after a photon passed the beam splitter B and before it could have reached point C, suddenly introduce a detector in the path s_2 at point C and do *not* detect *anything*, then it follows that the photon must have taken the path s_1—and, really, one can detect it in the region D, but there it does *not* produce interference any more ... The fact that by detecting *nothing* in point C we destroy the interference implies that the photon *somehow* knows of the other path when it takes the first one" [Pavičić, 1986, pp. 31, 32].

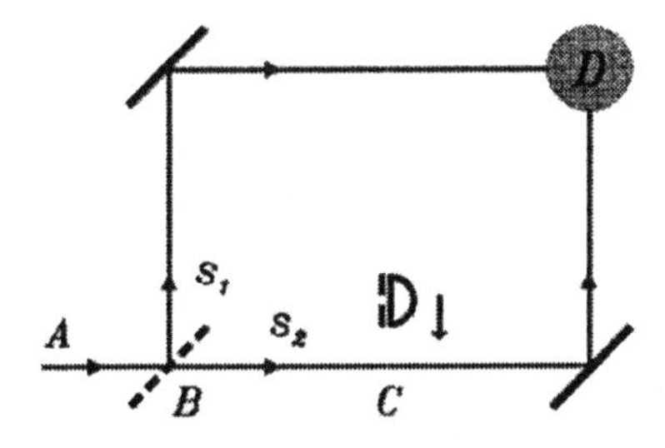

Figure 3.12. Figure taken from Pavičić (1986): "By detecting *nothing* in the point C we destroy the interference [in the region D]" [Pavičić, 1986, p.31]

This "photons's knowledge" is not explicitly described by the quantum mechanical formalism. The interference is described by the probability waves so, the aforementioned consideration of the paths of the energy carriers, i.e., of the photons themselves, was not unknown to physicists of the time but was considered useless. However, Elitzur and Vaidman [Elitzur and Vaidman, 1993] realized that path considerations of the waves can be used for realistic measurements, and several such experiments have been carried out since [Kwiat et al., 1999]. In what follows we will consider the experiments that enable a construction of an interaction-free CNOT gate. There are different proposals for a CNOT gate in the literature that use different interaction-free setups. For example, Gilchrist, White, and Munro arrive at a destructive *pseudo* CNOT gate in which one of the target qubits must be destroyed [Gilchrist et al., 2002]. On the other hand, Hiroo Azuma uses a positron as a straight moving object (control qubit) that blocks (N times, where $N \to \infty$) the paths of a zig-zag moving interrogating electron (target qubit) [Azuma, 2003; Azuma, 2004].

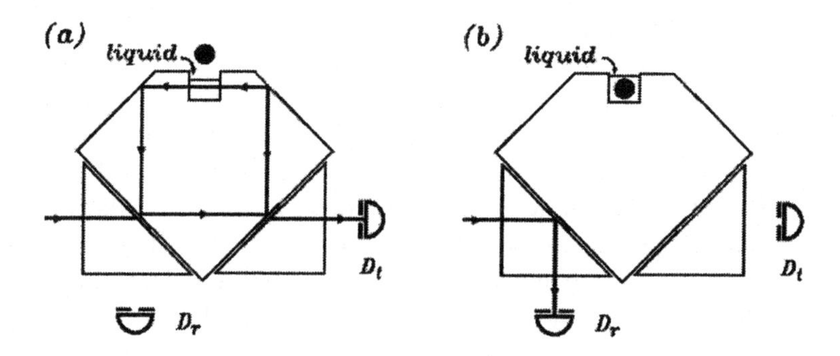

Figure 3.13. Figure taken from Paul and Pavičić (1997): "Lay-out of the proposed interaction-free experiment; (a) In the shown free round-trips the intensity of the reflected beam is approaching 0 for R approaching 1, i.e., detector D_r does not react; (b) However, when an absorbing object is immersed in the liquid (whose refractive index is the same as the one of the crystal in order to prevent losses of the free round-trips), for $R = 0.999$, 99.9% of the incoming beam reflect into D_r, 0.0001% go into D_t, and 0.0999% hit the object." [Paul and Pavičić, 1997]

Let us consider the setup proposed by Paul and Pavičić [Paul and Pavičić, 1996; Pavičić, 1996; Paul and Pavičić, 1997] and shown in Fig. 3.13. The outcomes have been confirmed by a real experiment carried out by Tsegaye, Goobar, Karlsson, Björk, Loh, and Lim (1998) [Tsegaye et al., 1998]. Our experimental proposal uses an uncoated monolithic total-internal-reflection resonator (MOTIRR) coupled to two triangular prisms by frustrated total internal reflection (FTIR). A squared MOTIRR requires a relative refractive index $n > 1.41$ with respect to the surrounding medium in order to confine a beam to the resonator (the angle of incidence being 45°). If, however, another medium (in our case, the right triangular prism in Fig. 3.13) is brought within a distance of the order of the wavelength, the total reflection within the resonator will be *frustrated* and a fraction of the beam will "tunnel out" from the resonator. Depending on the dimension of the gap and the polarization of the incidence beam one can well define reflectivity R within the range from 10^{-5} to 0.99995. The main advantage of such a coupling in comparison with coated resonators, is that the losses are extremely small—down to 0.003.

In the same way, a beam can "tunnel into" the resonator through the left triangular prism in Fig. 3.13, provided that the condition $n > 1.41$ is fulfilled for the prism too. The incident laser beam is chosen to be polarized perpendicularly to the incident plane so as to give a unique reflectivity for each photon. The faces of the resonator are polished spherically to give a large focusing factor. A round-trip path for the beam is created in the resonator, as shown in Fig. 3.13. A cavity is cut in the resonator and filled with an index-matching fluid to reduce losses. Now, if there is an object in the cavity in the round trip path of the beam in the resonator, the incident beam will be almost totally reflected (into D_r), and if there is no object, the beam will be almost totally transmitted (into D_t).

To understand this result, we sum up the contributions originating from round trips in the resonator to the reflected wave. The portion of the incoming beam of amplitude $A(\omega)$ reflected into the plane determined by δ is described by the amplitude $B_0(\omega) = -A(\omega)\sqrt{R}$, where $R = |r|^2$ is reflectivity. The transmitted part will travel around the resonator, guided by one frustrated total internal reflection (at the face next to the right prism) and by two proper total internal reflections. After a full round trip, the following portion of this beam joins the directly reflected portion of the beam by tunneling into the left prism: $B_1(\omega) = A(\omega)\sqrt{1-R}\sqrt{R}\sqrt{1-R}\, e^{i\psi}$. $B_2(\omega)$ contains three frustrated total internal reflections and so on; each subsequent round trip contributes to a geometric progression that gives the reflected amplitude

$$\begin{aligned} B_n(\omega) &= A(\omega)\sqrt{R}\{-1 + (1-R)e^{i\psi}[1 + Re^{i\psi} + (Re^{i\psi})^2 + \ldots]\} \\ &= \sum_{i=0}^{n} B_i(\omega)\,, \end{aligned} \tag{3.48}$$

where $\psi = (\omega - \omega_{res})T$ is the phase added by each round-trip. Here ω is the frequency of the incoming beam, T is the round trip time, and ω_{res} is the *selection frequency* corresponding to a wavelength that satisfies $\lambda = L/k$, where L is the round-trip length of the cavity and k is a positive integer. When summing up the round-trip contributions, we have taken into account that (because of the above condition imposed on the total phase shift ϕ) all contributions must lie in the reflected-wave plane and that their amplitudes must carry the opposite sign (to that of the reflected wave, $-A(\omega)\sqrt{R}$) so as to cancel out at resonance $\psi = 0$.

For an insight into the physics of the experiment, it is sufficient to consider plane waves ($A(\omega) = A_0$). The limit of $B_n(\omega)$ yields the total amplitude of the reflected beam:

$$B_r(\omega) = \lim_{n\to\infty} B_n(\omega) = -A_0\sqrt{R}\frac{1 - e^{i\psi}}{1 - Re^{i\psi}}\,. \tag{3.49}$$

We see that for any $R < 1$ and $\omega = \omega_{res}$, i.e., if nothing obstructs the round trip of the beam, we get no reflection at all (i.e., no response from D_r (see Fig. 3.13)). When an object blocks the round trip and R is close to 1, then we get almost a total reflection. In terms of single photons (which we can obtain by attenuating the intensity of a laser until the chance of having more than one photon at a time becomes negligible), the probability of detector D_r reacting when there is no object in the system is zero. A response from D_r means an interaction-free detection of an object in the system. The probability of the response is R, the probability of a photon hitting the object is $R(1-R)$, and the probability of a photon exiting into detector D_t is $(1-R)^2$. These results have been confirmed by several recent experiments [Tsegaye et al., 1998].

We can achieve a more realistic experimental approach by looking at two possible sources of individual photons, a CW laser (p. 33) and a pulse laser and by using wave packets instead of plane waves. [Pavičić, 1996; Paul and Pavičić, 1997; Paul and Pavičić, 1998; Pavičić, 2000b]. However, since all results as well as physics of the experiments remain the same under this more realistic approach, we will not enter into its details here.

There are two consequences of the above elaboration. One is instantaneous spreading of information to photons, and the other is the preparation of quantum states by means of information transfer—not energy transfer.

As for the first consequence, the switching on and off of the destructive interference of the waves does not spread within a finite interval of time by the speed of light but establishes propagation conditions for photons instantaneously. When we switch on a CW laser with a very low intensity, and a photon eventually comes to the device, it will "know" where to go. If an object is in the device, the photon will be reflected, and if not it will go through. Let us also look at the following experiment, shown in Fig. 3.14 and proposed by Fearn, Cook, and Milonni [Fearn et al., 1995].

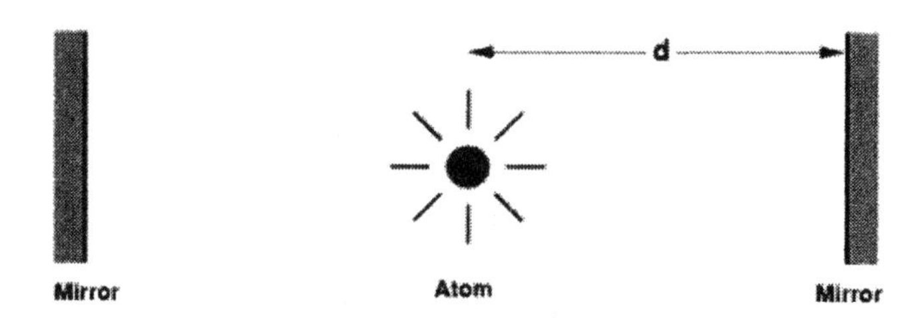

Figure 3.14. The figure follows Fearn, Cook, and Milonni (1995) [Fearn et al., 1995]: Inhibited photon emission of an excited atom in a cavity.

"Emission of an excited atom in a cavity is inhibited and the question is being addressed of whether a sudden replacement of one of the cavity mirrors by a detector can result in a photon count immediately or only after some retardation time ... [I]t is possible to count a photon immediately following the substitution of photodetector for a mirror" [Fearn et al., 1995]. This outcome has recently been confirmed by a real experiment with an inhibited down-conversion of photons from a crystal [Branning et al., 2003].

As for the other consequence, let us look at the experiment proposed by Pavičić [Pavičić, 1996] and shown in Fig. 3.15. It uses a combination of an atom interferometer with ultracold metastable atoms and resonance interaction-free path detection by means of a movable MOTIRR (of course, without liquid, which only slightly increases the efficiency). To increase the probability of an atom being hit by the round-trip beam, the incoming laser beam should be split into many beams by multiple beam splitters, each beam containing on average one photon in the chosen time window, so as to feed MOTIRR through many optical fibers. The atom source in the atom interferometer is a magneto-optical trap containing $1s_5$ neon metastable atoms which are then excited to the $2p_5$ state by a 598-nm laser beam. Of all the states to which $2p_5$ decays, we follow only $1s_3$ atoms, whose trajectories are determined only by the initial velocity and gravity (free fall from the trap).

Now the atoms fall with different velocities, but each velocity group forms interference fringes calculated as for the optical case and only corrected by a factor that arises from the acceleration by gravity during the fall. MOTIRR is mounted on a device that follows (with acceleration)

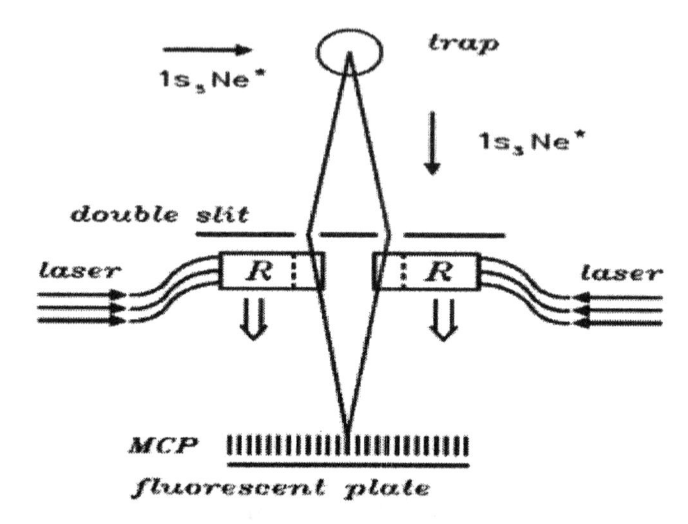

Figure 3.15. Figure taken from [Pavičić, 1996]: "Proposal for a *welcher Weg* atom interference experiment with ultracold atoms falling from a trap. MOTIRR resonators R, see Fig. 3.13, here shown sideways, move together with the falling atoms which sit in their openings" [Pavičić, 1996]

one velocity group from the double slit to the microchannel plate detector. (Atoms from other groups move with respect to MOTIRR and therefore cannot decohere MOTIRR.) The laser is tuned to a frequency equal to the $1s_3$ resonance frequency, which in effect increases the cross section of the atoms so as to make them efficiently "visible" to (virtual) photons. The source is attenuated so much that there is on average only one atom in a velocity group. The whole process repeats every 0.4 s. Assuming that we have 10 ns recovery time for the photon detectors and 300 optical fibers we arrive at about 10^7 counts, which all go into one detector D_t when no atom obstructs a round trip. As soon as detector D_r fires, we know which slit the observed atom passed through. After 10^3 repetitions of such successful detections, we have enough data to see that the interference fringes are destroyed significantly with respect to unmonitored reference samples.

Figuratively, one could call this device a "Heisenberg microscope without a kick." In it, the *welcher Weg* has been obtained without transferring a single bit of energy. "So, it is simply the information contained in a functioning measuring apparatus that changes the outcome of the experiment, and not uncontrollable alterations of the spatial wave function, resulting from the action of the measuring apparatus on the system under observation" [Scully et al., 1991].

To construct an interaction-free CNOT gate, we substitute an atom, for example ^{87}Rb (see the dark-state teleportation experiment in Sec. 3.1.6, p. 146) for the object and liquid in our MOTIRR in Fig. 3.13 (p. 160). We put the atom from Fig. 3.7 (p. 150) into the gap of our resonator in Fig. 3.13 (p. 160). The ^{87}Rb atom will be transparent for properly polarized photons of definite frequency when there is no electron in the ground level that a photon could excite to a higher level (a photon will not "see" the atom—see p. 111), and nontransparent when there is.

Looking at Fig. 3.6 (p. 148) and the states on p. 149, we see that the left-hand circularly polarized photon can excite the atom from its ground state $|g_1\rangle$ ($5s_{1/2}$, $F = 1$, $m = -1$) to its excited state $|e\rangle$ ($5p_{1/2}$, $F = 2$, $m = 0$), and the right circularly polarized photon can excite the atom from $|g_2\rangle$ ($5s_{1/2}$, $F = 1$, $m = +1$) to $|e\rangle$ ($5p_{1/2}$, $F = 2$, $m = 0$) too. So an L-photon will "see" the atom in $|g_1\rangle$ but will not "see" it when it is in the state $|g_2\rangle$. With an R-photon, it is the other way around. The energy differences to the detuned excited level are the same, so both photons have the same frequency. We can induce a change of the atom from $|g_1\rangle$ to $|g_2\rangle$ and back by a STIRAP process, with two additional external laser beams, as explained in Sec. 3.1.4 (p. 142).

We feed our resonator with $+45°$ and $-45°$ linearly polarized photons to achieve the same conditions for both kinds of photons. At the right-hand side of the gap, we put a quarter-wave plate, QWP (see p. 22) to turn a $45°$-photon into an R-photon and a $-45°$-photon into an L-photon. At the exit from the gap, we direct the photons back into the resonator through a half-wave plate, HWP (see p. 22), to change the direction of the circular polarization and into another QWP to transform it back into the original linear polarization. We denote the atom states as follows:

$$|0\rangle = |g_1\rangle, \qquad |1\rangle = |g_2\rangle, \tag{3.50}$$

and will take these atom states to be the control states and the atom itself our control qubit. We denote the photon states as follows:

$$|0\rangle = |45°\rangle, \qquad |1\rangle = |-45°\rangle, \tag{3.51}$$

and will take these photon states as the target states and photons as target qubits. For example, $|01\rangle$ means that the atom is in the $g1$ state and the photon is polarized along $-45°$.

Let us convince ourselves that we really have a CNOT gate. Consider first Fig. 3.16 (a). A photon in the state $|0\rangle$ does not "see" the atom in the state $g1$ and will therefore exit the resonator through the right port and will pass HWP. At HWP its state turns into the state $|1\rangle$ and reflects at the polarizing beam splitter PBS in Fig. 3.16. A photon in

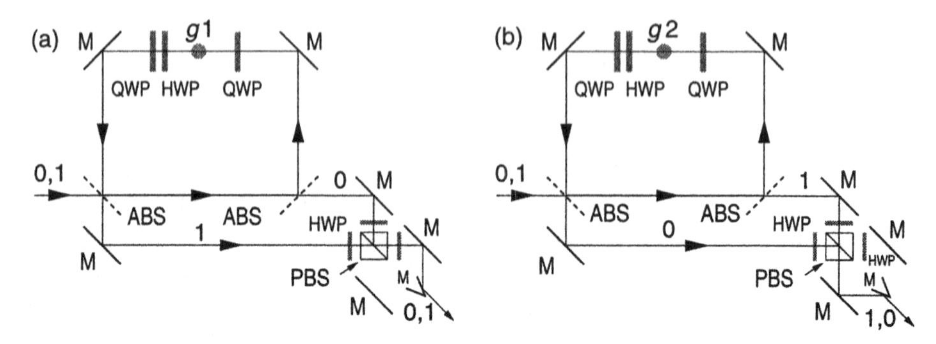

Figure 3.16. Interaction-free CNOT. (a) The atom is in the $g1$ state and can absorb a 45° polarized photon 1. Therefore, photon 1 (original photon) in state $|1\rangle$) cannot enter the cavity, and we have $0 \to 0$ and $1 \to 1$. (b) The atom is in the $g1$ state and can absorb a $-45°$ polarized photon 0. Therefore photon 0 cannot enter the cavity, and we have $0 \to 1$ and $1 \to 0$. ASM are highly asymmetrical beam splitters with $R = 0.999$; M are perfect mirrors; HWP and QWP are half- and quarter-wave plates, respectively and PBS is a polarizing beam splitter which lets 0 photons through and reflects 1 photons. QWP, HWP, and QWP in the resonator turn linear polarization into circular and back into linear.

the state $|1\rangle$ "sees" the atom in the state $g1$ and therefore does not enter the resonator but goes down to HWP, changes its state into $|0\rangle$, passes PBS,0 and goes to HWP, which changes its state back to $|1\rangle$. Fig. 3.16 (b) refers to the atom in state $|1\rangle$. A photon in state $|0\rangle$ sees it, goes down to HWP, turns into $|1\rangle$, and PBS reflects it downward. A $|1\rangle$ photon does not see the atom and exits through the right port, changes the polarization at PBS into $|0\rangle$, and passes PBS. Hence we have

$$|00\rangle \to |00\rangle, \qquad |01\rangle \to |01\rangle, \qquad |10\rangle \to |11\rangle, \qquad |11\rangle \to |10\rangle, \tag{3.52}$$

which was to be verified.

In this way, we obtain not only a robust interaction-free CNOT gate but also a nondestructive way to detect the states of an atom. This approach promises an integration of interaction-free circuits into would-be quantum computers. Conditions under which we can make extensive interaction-free computations have been considered by Mitchison and Jozsa (under the name *counterfactual computation*) [Mitchison and Jozsa, 2001].

3.2.2 Kochen–Specker Setups

As we have repeatedly seen throughout this book, the most important nonclassical "ingredients" of quantum formalism are superposition and entanglement.[10] And yet, all the measurements that can possibly be made of quantum systems are "binary": everything reduces to having (1) or not having (0) a click in a detector. It is well known that any attempt to make a classical theory (often called a *hidden variable theory*) of these binary outputs from quantum measurements would fail. It is, however, important to know whether there is a set of measurements that can be carried out on a finite dimensional quantum system in such a way that if one assumed that the values of measured observables are completely independent of all other observables that can be measured on the same system, then one would run into a contradiction. The so-called Kochen–Specker theorem [Kochen and Specker, 1967] answers this question in the affirmative and proves that a quantum system cannot possess a definite value of a measurable property prior to measurement, and that quantum measurements (essentially detector clicks) carried out on quantum systems cannot be ascribed predetermined values (say 0 and 1). The next knowledge we would like to have about such systems is how often we can encounter them and whether we can use them for setups that would verify whether a given theory is classical or quantum, for example, an algebra that one would like to use for a future quantum computer. Such setups we will call *KS setups.*[11]

Recently, algorithms have been put forward that can in principle generate all such systems up to a reasonable number of observables, dimensions, and experimental conditions [Pavičić et al., 2005]. In this section we will present these algorithms, the resulting setups, and experiments that has verified them.

Let us consider an orthonormal set of states $\{\psi_1, \dots, \psi_n\}$, i.e., vectors in n-dimensional Hilbert space, $\mathcal{H}^n$, $n \geq 3$. Projectors onto these states satisfy: $\sum_{i=1}^{n} P_i = I$, where $P_i = \psi_i \psi_i^\dagger$. Now, Kochen and Specker proved [Kochen and Specker, 1967] that there is no function $f : \mathcal{H} \rightarrow \mathbb{R}$ satisfying the Sum Rule $\sum_{i=1}^{n} f(P_i) = f(\sum_{i=1}^{n} P_i) = f(I)$ for all sets of projectors P_i. Hence, there is at least one set of projectors $\{P_i, P_i', \dots\}$

[10]Quantum entanglement emerges as a consequence of our free engineering of quantum states. Some tasks for which we usually employ entanglement can be carried out without it. But avoiding entanglement does not seem to speed up solving the task. Whether entanglement is theoretically necessary for quantum computation speedup is an open problem.

[11]The essential difference between the Kochen–Specker setup and a setup used to verify the so-called *Bell inequalities* is that for the some arrangements of the latter setup, the quantum and classical results agree for one group of measurements and disagree for another, while for the former setup they always disagree.

and the corresponding set of vectors $\{\psi_i, \psi'_i, \ldots\}$ for which the Sum Rule is not satisfied. If we choose $f(P_i) \in \{0,1\}$ $(f(I) = 1)$, the theorem amounts to the following claim: In $\mathcal{H}^n$, $n \geq 3$, it is impossible to assign 1s and 0s to all vectors from such a set—which we call a *KS (Kochen–Specker) set*—in such a way that [Zimba and Penrose, 1993]:

1. No two orthogonal vectors are both assigned the value 1; and
2. In any subset of n mutually orthogonal vectors, not all the vectors are assigned the value 0.

All the vectors from a KS (Kochen–Specker) set, as defined above, we call *KS (Kochen–Specker) vectors* [Pavičić et al., 2005]. KS vectors in each KS set form subsets of n mutually orthogonal vectors. We arrive at one subset from another by a series of rotation in 2-dimensional planes around $(n-2)$-dimensional subspaces. Thus, any two subsets share at least one vector that is orthogonal to all other vectors in both subsets, and in an n-dimensional space, two subsets can share up to $n-2$ vectors. The KS vectors correspond to the directions of the quantization axes of the measured eigenstates within experiments that have no classical interpretation, and when we speak of finding KS vectors we mean finding these directions. We emphasize here that it is not our aim to give yet another proof of the KS theorem but rather to determine the class of all KS vectors from an arbitrary $\mathcal{H}^n$ as well as the class of all *non-KS vectors*, i.e., vectors from the remaining sets of vectors from $\mathcal{H}^n$. By the class of non-KS vectors, we mean vectors that allow 0–1 states and that correspond to the directions of the quantization axes of the measured eigenstates within experiments that do have classical interpretation, and when we speak of finding non-KS vectors we mean finding the latter directions.

This approach is based on a recognition that a description of a discrete observable measurement (e.g., spin) in $\mathcal{H}^n$ can be rendered as a 0–1 measurement of the corresponding projector along the vector in $\mathbb{R}^n$ onto which the projector projects. Hence, we deal with orthogonal triples in $\mathbb{R}^3$, quadruples in $\mathbb{R}^4$, etc., which correspond to possible experimental designs, and to find KS vectors means finding such n-tuples in $\mathbb{R}^n$.

The algorithms (*Kochen–Specker (KS) algorithms*) for finding these n-tuples, i.e., KS vectors, start with the so-called MMP diagrams [Pavičić et al., 2005; Pavičić, 2002; McKay et al., 2000]. Eventually, we will pick up a special and very small class of MMP diagrams whose vertices (points) will correspond to KS vectors and whose edges (lines connecting vertices) will correspond to orthogonalities between KS vectors. But the definition of MMP diagrams is much more general:

DEFINITION 3.2 *MMP diagrams are diagrams whose vertices and edges satisfy the following conditions:*

1. *Every vertex belongs to at least one edge;*

2. *Every edge contains at least three vertices;*

3. *Edges that intersect with each other in $n-2$ vertices contain at least n vertices;*

We generate MMP diagrams by using the so-called *isomorphism-free generation* that follows the general principles established by [McKay, 1998], which can summarized as follows.

Deleting an edge from an MMP diagram, together with any vertices that lie only on that edge, yields another MMP diagram (perhaps the vacuous one with no vertices). Consequently, every MMP diagram can be constructed by starting with the vacuous diagram and adding one edge at a time, at each stage having an MMP diagram.

We can represent this process as a rooted tree whose vertices correspond to MMP diagrams whose vertices and edges have unique labels. The vacuous diagram is at the root of the tree, and for any other diagram its parent node is the diagram formed by deleting the edge with the highest label. The isomorph rejection problem is to prune this tree until it contains just one representative of each isomorphism class of diagram. This result can be achieved by the application of two *rules.*

Given a diagram D, we can identify the valid positions to add a new edge such that Conditions 2–3 are enforced. According to the symmetries of D, some of these positions are equivalent. The first rule is that exactly one position in each equivalence class of positions is used; a node in the tree formed by adding an edge in any other position is deleted together with all its descendants.

To understand the second rule, consider a diagram D' with at least one edge. We label the edges of D' in a canonical order, which is an order independent of any previous labeling. Then we define the *major class* of edges as those that are equivalent under the symmetries of D' to the edge that is last in canonical order. The second rule is: when D' is constructed by adding an edge e to a smaller diagram, delete D' (and all its descendants) unless e is in the major class of edges of D'.

According to the theory in [McKay, 1998], the application of both rules together is sufficient: exactly one diagram from each isomorphism class remains in the tree. A generation tree for MMP diagrams with nine vertices and the smallest loop of size 5 is shown in Fig. 3.17.

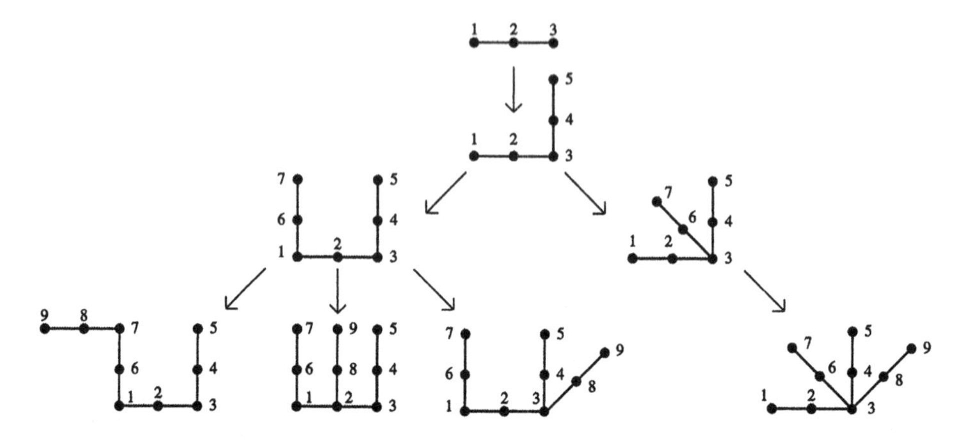

Figure 3.17. An example of a generation tree for connected MMP diagrams: nine vertices and the smallest loop of size 5 (for nine vertices, a loop cannot be formed; the first loop appears with 10 vertices: `123,345,567,789,9A1`—see text) [McKay et al., 2000; Pavičić et al., 2005].

We denote vertices of MMP diagrams by `1,2,..,A,B,..a,b,..` By the above algorithm, we generate MMP diagrams with chosen numbers of vertices and edges and a chosen minimal loop size.

Now we want to find diagrams that cannot be ascribed 0–1 values. To this end, we apply an algorithm for an exhaustive search of MMP diagrams with backtracking. The criterion for assigning 0–1 (dispersion-free) states is that each edge must contain exactly one vertex assigned to 1, with the others assigned to 0. As soon as a vertex on an edge is assigned a 1, all other vertices on that edge become constrained to 0, and so on. The algorithm scans the vertices in some order, trying 0 then 1, skipping vertices constrained by an earlier assignment. When no assignment becomes possible, the algorithm backtracks until all possible assignments are exhausted (no solution) or a valid assignment is found. For the investigated diagrams, it has been found that the average time per diagram grows polynomially with the diagram size.

The obtained diagrams correspond to candidate sets of nonlinear equations that contain KS sets in the following sense. The number of KS vectors, i.e., the number of vertices within edges, corresponds to the dimension of $\mathbb{R}^n$, and the edges in turn correspond to $n(n-1)/2$ equations resulting from inner products of vectors being equal to zero, which means orthogonality. So, e.g., an edge of length 3, `BCD`, represents the following three equations:

$$\begin{aligned}\mathbf{a}_B \cdot \mathbf{a}_C &= a_{B1}a_{C1} + a_{B2}a_{C2} + a_{B3}a_{C3} &= 0,\\ \mathbf{a}_B \cdot \mathbf{a}_D &= a_{B1}a_{D1} + a_{B2}a_{D2} + a_{B3}a_{D3} &= 0,\\ \mathbf{a}_C \cdot \mathbf{a}_D &= a_{C1}a_{D1} + a_{C2}a_{D2} + a_{C3}a_{D3} &= 0.\end{aligned} \qquad (3.53)$$

Each possible combination of edges for a chosen number of vertices corresponds to a system of such nonlinear equations. A solution to systems that correspond to MMP diagrams without 0–1 states is a set of components of KS vectors that we want to find. Thus a direct approach would be an attempt to find *all* KS vectors in the exhaustive generation of all MMP diagrams, then pick out all those diagrams that cannot have 0–1 states, establish the correspondence between the latter diagrams and the equations for the vectors as shown in Eq. (3.53), and finally solve the systems of the equations so obtained. However, there is a no-go reason for such an approach: for the smallest KS set of 18 KS vectors we should generate $> 2.9 \times 10^{16}$ systems, and that would require more than 30 million years on a 2 GHz CPU. For higher KS sets, this time grows exponentially.

The KS algorithms get around the time barrier so as to merge the above stages. They first generate only those diagrams that cannot have a solution. This is crucial. Without this, we would not be able to reduce the exponential complexity of the problem to the statistically polynomial one. For systems of equations of the type given by Eq. (3.53) that do have solutions that do not allow 0–1 states, such solutions are KS vectors that correspond to vertices of MMP diagrams.

To find these solutions to nonlinear equations of the type given by Eq. (3.53) is again a highly nontrivial task. Therefore, algorithms that reduce the task complexity to a statistically polynomial one based on interval analysis and Ritt's characteristic set calculations have been developed [Pavičić et al., 2005]. This approach rounds up the constructive and exhaustive definition of KS sets and vectors and makes their generation feasible for reasonably chosen numbers of vectors and dimensions.

Applying these to, say, $\mathbb{R}^4$, we can obtain the results shown in Figs. 3.18 and 3.19. When we take into account that there are billions and billions of equations that are candidates for just one smallest KS set of 18 KS vectors, it might seem astonishing that these systems have previously been found by "humans" Cabello, Estebaranz, and Guillermo [Cabello et al., 1996]—although they could not have known whether their system was the smallest one. Actually, it shows that we are increasingly able to find shortcuts for handling quantum phenomena.

Let us consider the experiment proposed by Cabello and García-Alcaine [Cabello and García-Alcaine, 1998], which has been experimen-

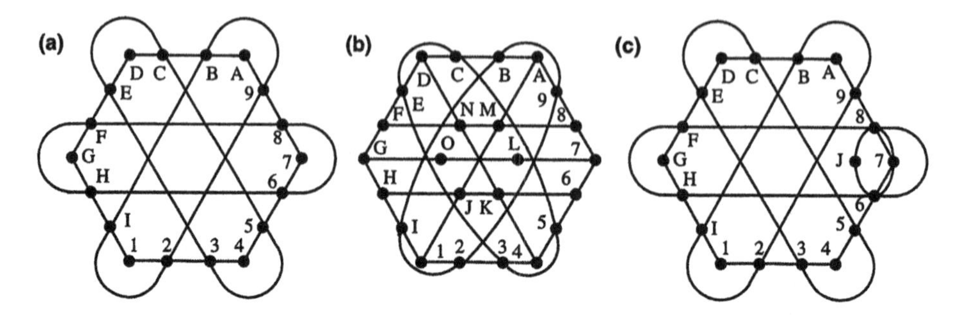

Figure 3.18. The smallest 4-dimensional KS systems with: (1) loops of size 3: (a) 18-9 (isomorphic to Cabello et al. [Cabello et al., 1996]); (b) 24(22)–13 not containing system (a), with values $\notin \{-1, 0, 1\}$; (2) loops of size 2: (c) 19(18)–10. Between systems (a) and (b) there are 62 systems with loops of size 3, all containing the system (a).

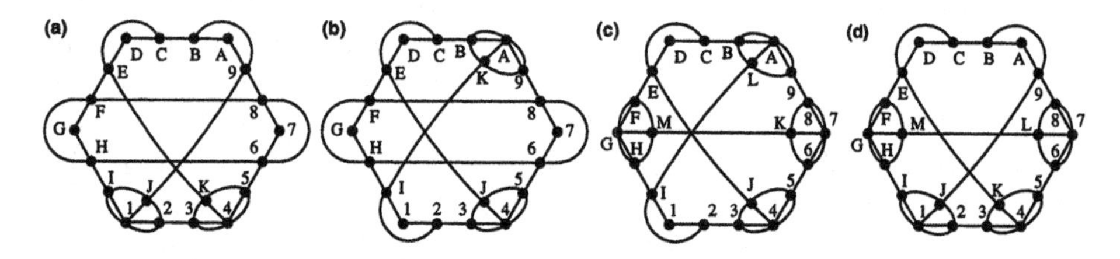

Figure 3.19. Smallest 4-dimensional KS systems with loops of size 2: (1) — not containing system (a) of Fig. 3.18: (a) 20–11; (b) 20–11 isomorphic to Kernaghan [Kernaghan, 1994]; (2) — containing neither system (a) of Fig. 3.18 nor systems (a) and (b) of this figure: (c) 22–13; (d) 22–13.

tally confirmed [Huang et al., 2002]. The experiment realizes a two-qubit ($\mathbb{R}^4 = \mathbb{R}^2 \otimes \mathbb{R}^2$) photon experiment, where the first qubit is represented by the photon path and the second qubit by the photon polarization. We prepare the following single-photon two-qubit state

$$|\Psi\rangle = \frac{1}{\sqrt{2}}(|00\rangle + |11\rangle), \tag{3.54}$$

where $|00\rangle$ means $|0\rangle_1|0\rangle_2$, where $|0\rangle_1$ means a horizontally polarized photon, $|1\rangle_1$ a vertically polarized photon, $|0\rangle_2$ a photon going through a polarizing beam splitter, and $|1\rangle_2$ a photon being reflected at a polarizing beam splitter. Our operators are

$$\begin{aligned} Z_1 &= |0\rangle_{1\,1}\langle 0| - |1\rangle_{1\,1}\langle 1|, & X_1 &= |0'\rangle_{1\,1}\langle 0'| - |1'\rangle_{1\,1}\langle 1'|, \\ Z_2 &= |0\rangle_{2\,2}\langle 0| - |1\rangle_{2\,2}\langle 1|, & X_2 &= |0'\rangle_{2\,2}\langle 0'| - |1'\rangle_{2\,2}\langle 1'|, \end{aligned} \tag{3.55}$$

where

$$|0'\rangle_i = \frac{1}{\sqrt{2}}(|0\rangle_i + |1\rangle_i), \qquad |1'\rangle_i = \frac{1}{\sqrt{2}}(|0\rangle_i - |1\rangle_i), \qquad i = 1, 2. \quad (3.56)$$

One performs the experiment measuring various X_i, Z_i combinations according to a table given by Adán Cabello [Cabello, 2000], which corresponds to the KS set shown in Fig. 3.18 (a). Quantum mechanics, of course, always gives a definite set of results, but an attempt to keep the outcomes of previous measurements fixed and to use them in subsequent ones must fail. If we try to ascribe 0 and 1 to the points in Fig. 3.18 (a) according to the rules from p. 168 (any edge must contain one 1 and three 0s), we will soon find that this is impossible.

3.3 Quantum Algorithms

As we mentioned on pp. xiii and 31, in 1947 the head of that computing center in Harvard, Howard Aiken estimated the no more than six computers would satisfy the computing needs of the entire United States. One reason for such an underestimate of future computing needs was the absence of Boolean algorithms and software at the time. Today we have a similar situation with quantum algorithms and quantum software for would-be quantum computers. Practically all known quantum algorithms are based on a single function—the *quantum Fourier transform* (a quantum version of the classical *discrete Fourier transform*). On the other hand, there is still no universal quantum algebra for quantum computers analogous to Boolean algebra for classical computers. Therefore, we will present several algorithms and in the end discuss possibilities for constructing a universal quantum algebra.

3.3.1 Quantum Coin—Deutsch's Algorithm

When a magician performs a trick with a classical coin, we can only see the top side of it, which will show either heads or tails. We are curious to learn whether the coin is fair or fake (having heads on both sides or tails on both sides), but we are not allowed to climb the stage to turn the coin over and look at the bottom side. However, if we gave the magician a quantum coin and used what is known as *Deutsch's algorithm* [Deutsch, 1985], we would be able to distinguish a fair from a fake coin in one step.

The algorithm uses two kinds of *evaluating functions* $f\colon \{0,1\} \to \{0,1\}$:

- *Constant functions* $f_1(x) = 0$ and $f_2(x) = 1$, where 0 and 1 stand for heads and tails respectively (or the other way around) and x for the coin's top side that the magician shows us ($x = 0$) and the bottom side that we are classically not allowed to check ($x = 1$) (or the other way around). Constant functions describe fake coins.

- *Balanced functions* $f_3(0) = 0,\ f_3(1) = 1$ and $f_4(0) = 1, f_4(1) = 0$, meaning that the top side shows heads while the bottom side is tails and that the top shows tails while the bottom is heads, respectively. Balanced functions describe fair coins.

Since our goal is to learn whether the coin is fair or fake, we will dispense with any information about whether the coin shows heads or tails at a particular time and will deal only with a superposition of the top and bottom sides of the coin and with a superposition of heads and tails. We define the following gate, controlled by the function f, and let it act on a superposition of states. The final measurement will reveal whether f is constant or balanced.

DEFINITION 3.3 *An f–CNOT gate is defined as*

$$|x\rangle|y\rangle \xrightarrow{f-\text{CNOT}} |x\rangle|y \oplus f(x)\rangle, \tag{3.57}$$

where $\oplus$ is the XOR *operation (addition modulo 2—see p. 8).*

The constant functions leave the state of the first qubit unchanged up to a sign:

$$\begin{aligned}
|x_{\text{in}}\rangle|y\rangle &= \frac{1}{\sqrt{2}}(|0\rangle+|1\rangle)\frac{1}{\sqrt{2}}(|0\rangle-|1\rangle) = \frac{1}{2}(|0\rangle|0\rangle-|0\rangle|1\rangle+|1\rangle|0\rangle-|1\rangle|1\rangle)\\
&\xrightarrow{f_1} \frac{1}{2}(|0\rangle|0\oplus f_1(0)\rangle - |0\rangle|1\oplus f_1(0)\rangle\\
&\qquad +|1\rangle|0\oplus f_1(1)\rangle - |1\rangle|1\oplus f_1(1)\rangle)\rangle\\
&= \frac{1}{2}(|0\rangle|0\rangle - |0\rangle|1\rangle + |1\rangle|0\rangle - |1\rangle|1\rangle)\\
&= |x_{\text{in}}\rangle|y\rangle \qquad \Rightarrow \qquad |x_{\text{out}}\rangle = |x_{\text{in}}\rangle,\\
|x_{\text{in}}\rangle|y\rangle &\xrightarrow{f_2} \frac{1}{2}(|0\rangle|0\oplus f_2(0)\rangle - |0\rangle|1\oplus f_2(0)\rangle\\
&\qquad +|1\rangle|0\oplus f_2(1)\rangle - |1\rangle|1\oplus f_2(1)\rangle)\rangle\\
&= \frac{1}{2}(|0\rangle|1\rangle - |0\rangle|0\rangle + |1\rangle|1\rangle - |1\rangle|0\rangle)\\
&= -|x_{\text{in}}\rangle|y\rangle \qquad \Rightarrow \qquad |x_{\text{out}}\rangle = -|x_{\text{in}}\rangle,
\end{aligned} \tag{3.58}$$

while the balanced functions change the superposition of the first qubit state:

$$
\begin{aligned}
|x_{\rm in}\rangle|y\rangle &\xrightarrow{f_3} \frac{1}{2}(|0\rangle|0\oplus f_3(0)\rangle - |0\rangle|1\oplus f_3(0)\rangle \\
&\qquad +|1\rangle|0\oplus f_3(1)\rangle - |1\rangle|1\oplus f_3(1)\rangle)) \\
&= \frac{1}{2}(|0\rangle|0\rangle - |0\rangle|1\rangle + |1\rangle|1\rangle - |1\rangle|0\rangle) \\
&= \frac{1}{\sqrt{2}}(|0\rangle - |1\rangle))|y\rangle \qquad \Rightarrow \qquad |x_{\rm out}\rangle = \frac{1}{\sqrt{2}}(|0\rangle - |1\rangle), \\
|x_{\rm in}\rangle|y\rangle &\xrightarrow{f_4} \frac{1}{2}(|0\rangle|0\oplus f_4(0)\rangle - |0\rangle|1\oplus f_4(0)\rangle \\
&\qquad +|1\rangle|0\oplus f_4(1)\rangle - |1\rangle|1\oplus f_4(1)\rangle)) \\
&= \frac{1}{2}(|0\rangle|1\rangle - |0\rangle|0\rangle + |1\rangle|0\rangle - |1\rangle|1\rangle) \\
&= -\frac{1}{\sqrt{2}}(|0\rangle - |1\rangle))|y\rangle \;\Rightarrow\; |x_{\rm out}\rangle = -\frac{1}{\sqrt{2}}(|0\rangle - |1\rangle). \quad (3.59)
\end{aligned}
$$

By applying the Hadamard gate (see p. 32) to $|x_{\rm out}\rangle$'s given by Eq. (3.58), we get

$$
\hat{\rm H}\frac{\pm 1}{\sqrt{2}}(|0\rangle + |1\rangle) = \pm\frac{1}{2}\begin{bmatrix} 1 & 1 \\ 1 & -1 \end{bmatrix}\begin{bmatrix} 1 \\ 1 \end{bmatrix} = \pm\begin{bmatrix} 1 \\ 0 \end{bmatrix} = \pm|0\rangle. \qquad (3.60)
$$

Applying it to $|x_{\rm out}\rangle$'s given by Eq. (3.59) yields

$$
\hat{\rm H}\frac{\pm 1}{\sqrt{2}}(|0\rangle - |1\rangle) = \pm\frac{1}{2}\begin{bmatrix} 1 & 1 \\ 1 & -1 \end{bmatrix}\begin{bmatrix} 1 \\ -1 \end{bmatrix} = \pm\begin{bmatrix} 0 \\ 1 \end{bmatrix} = \pm|1\rangle. \qquad (3.61)
$$

Therefore, the measurement presented in Fig. 3.20 distinguishes constant (f_1, f_2) from balanced (f_3, f_4) functions in one step.

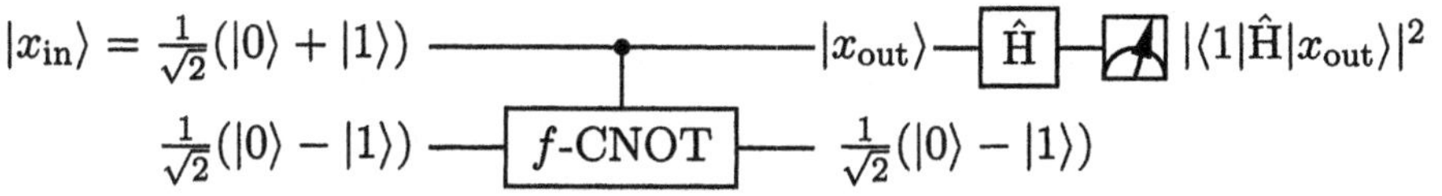

$|x_{\rm in}\rangle = \frac{1}{\sqrt{2}}(|0\rangle + |1\rangle)$ —•— $|x_{\rm out}\rangle$ — $\hat{\rm H}$ — measurement — $|\langle 1|\hat{\rm H}|x_{\rm out}\rangle|^2$

$\frac{1}{\sqrt{2}}(|0\rangle - |1\rangle)$ — f-CNOT — $\frac{1}{\sqrt{2}}(|0\rangle - |1\rangle)$

Figure 3.20. The f-CNOT gate of Deutsch's algorithm. The measurements of $|x_{\rm out}\rangle$ obtained by constant function gates f_1-CNOT and f_2-CNOT [see Eq. (3.58)] yield $|\langle 1|\hat{\rm H}|x_{\rm out}\rangle|^2 = 0$ [see Eq. (3.60)], and the ones carried out by balanced function gates f_3-CNOT and f_4-CNOT [see Eq. (3.59)] yield $|\langle 1|\hat{\rm H}|x_{\rm out}\rangle|^2 = 1$ [see Eq. (3.61)].

Successful implementations of the Deutch algorithm have been carried out on NMR [Dorai et al., 2000] and ion-trap [Gulde et al., 2003] quantum computers. These implementations relied on CNOT gates (see pp. 98 and 122). We will not repeat them here but would like to stress an important difference between CNOT and f-CNOT gates. CNOT gates give different outputs for different inputs of the qubit states. With f-CNOT gates, the input states stay the same, while different f-CNOT gates (depending on function f) give different outputs. Different f-CNOT gates have been implemented by different by NMR spin-selective pulses or by laser pulses that drive the electron (qubit) transitions within an ion-trap computer. How to make f-CNOT gates become part of active quantum circuits and play a role in quantum computation remains an open question.

For a generalization of Deutsch's algorithm in the next section, we rewrite Eqs. (3.58) and (3.59) in the following condensed form (see [Cleve et al., 1998])

$$\begin{aligned} |x_{\text{out}}\rangle &= \frac{1}{\sqrt{2}}((-1)^{f(0)}|0\rangle + (-1)^{f(1)})|1\rangle \\ &= \frac{(-1)^{f(0)}}{\sqrt{2}}(|0\rangle + (-1)^{f(0)\oplus f(1)}|1\rangle), \end{aligned} \tag{3.62}$$

where we have assumed that $|x_{in}\rangle = (|0\rangle + |1\rangle)/\sqrt{2}$ and f covers all four f_i, $i = 1, \ldots, 4$. The application of the Hadamard transform given by Eqs. (3.60) and (3.61) can be written as (see [Cleve et al., 1998])

$$\hat{\text{H}}|x_{\text{out}}\rangle = (-1)^{f(0)}|f(0) \oplus f(1)\rangle. \tag{3.63}$$

Note that $|x\rangle$ and $|y\rangle$ are not detached from each other as the final outcomes of Eqs. (3.58), (3.59), and (3.62) might suggest at a first glance—Eq. (3.62) is just a handy way to write down the x-qubit part of the right-hand side of Eq. (3.57). Also, we can start with $|x_{\text{in}}\rangle = |0\rangle$ and obtain the initial superposition by using a Hadamard gate: $\hat{\text{H}}|0\rangle = (|0\rangle + |1\rangle)/\sqrt{2}$.

3.3.2 Deutsch-Jozsa and Bernstein-Vazirani Algorithms

A generalization of Deutsch's algorithm to functions $f : \{0,1\}^n \to \{0,1\}$ was given by Deutsch and Jozsa [Deutsch and Jozsa, 1992] and simplified by Cleve, Ekert, Macchiavello, and Mosca [Cleve et al., 1998]. In the *Deutsch-Jozsa algorithm*, f is "promised" to be either constant (2^n values are all either equal to 0 or to 1) or balanced (exactly half, $2^n/2 = 2^{n-1}$ of the values are equal to 0 and the other half to 1), and we have to find out which one is given to us.

The f-CNOT gate definition stays the same as in Def. 3.3, except that now $x \in \{0,1\}^n$. For example, for $n = 2$ we have four possible values for $|x\rangle$:

$$|0\rangle|0\rangle = |00\rangle, \quad |0\rangle|1\rangle = |01\rangle, \quad |1\rangle|0\rangle = |10\rangle, \quad \text{and} \quad |1\rangle|1\rangle = |11\rangle. \tag{3.64}$$

The corresponding $|y \oplus f(x)\rangle$ from Def. 3.3 then yields:

$$\begin{aligned} &|y \oplus f(0) \oplus f(0)\rangle, \qquad |y \oplus f(0) \oplus f(1)\rangle, \\ &|y \oplus f(1) \oplus f(0)\rangle, \quad \text{and} \quad |y \oplus f(1) \oplus f(1)\rangle, \end{aligned} \tag{3.65}$$

Our input state is

$$|x_{\text{in}}\rangle = |0\rangle|0\rangle \ldots |0\rangle \frac{1}{\sqrt{2}}(|0\rangle - |1\rangle) = |00\ldots 0\rangle \frac{1}{\sqrt{2}}(|0\rangle - |1\rangle). \tag{3.66}$$

By applying the Hadamard gate on each state of n qubits, we get

$$\begin{aligned} \hat{\mathrm{H}}^{(n)}|x_{\text{in}}\rangle &= \hat{\mathrm{H}}|0\rangle\hat{\mathrm{H}}|0\rangle \ldots \hat{\mathrm{H}}|0\rangle \frac{1}{\sqrt{2}}(|0\rangle - |1\rangle) \\ &= \frac{1}{\sqrt{2^{n+1}}}(|0\rangle + |1\rangle)(|0\rangle + |1\rangle)\ldots(|0\rangle + |1\rangle)(|0\rangle - |1\rangle) \\ &= \frac{1}{\sqrt{2^{n+1}}}\sum_{j=0}^{2^n-1}|j\rangle(|0\rangle - |1\rangle), \end{aligned} \tag{3.67}$$

where j's denote the values in the binary expansion of $2^n - 1$ as given by Eq. (1.65), (1.66), and (1.67) on p. 37. For example, for $n = 2$, j's are given by Eq. (3.64).

The next step is to apply f-CNOT to the obtained states (given by Eq. (3.67)) as represented in Fig. 3.21. This boils down to applying f-CNOT to each qubit simultaneously using Eq. (3.62):

$$\frac{1}{\sqrt{2^{n+1}}}\sum_{j=0}^{2^n-1}(-1)^{f(j)}|j\rangle(|0\rangle - |1\rangle). \tag{3.68}$$

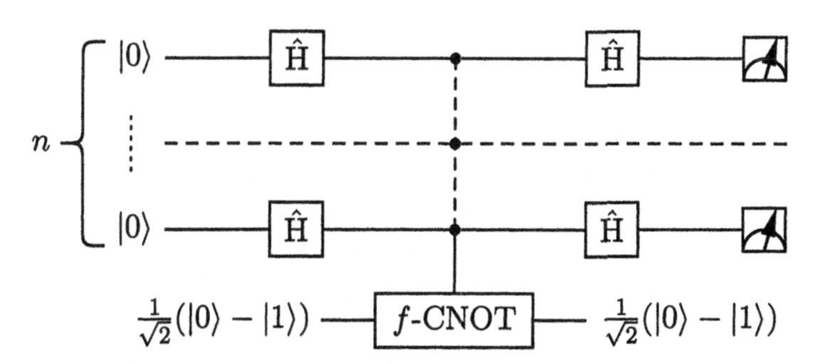

Figure 3.21. Circuit diagram for Deutch-Jozsa and Bernstein-Vazirani (p. 179) algorithms.

Now, let us estimate the number of times we would have to call the function f under the assumption of fair sampling, i.e., assuming that we never get the same display of values (binary input) twice—this is a stricter requirement than just randomly choosing binary values. Simple combinatorial reasoning (iterating through the possible combinations of 0s and 1s) reveals that we have to call function f exactly $2^{n-1}+1$ times to find out whether f is constant or balanced with *certainty*. For example, for $n=3$ and $y=0$ in Def. 3.3 we can have the following x's as inputs: 000, 101, 110, and 011. They can give (the first 0 is y): $0\otimes0\otimes0\otimes0=0$, $0\otimes1\otimes0\otimes1=0$, $0\otimes1\otimes1\otimes0=0$, and $0\otimes0\otimes1\otimes1=0$. We see that both constant ($f(0)=f(1)=0$) and balanced ($f(0)=0,\ f(1)=1$) functions give the same result and that the first next test ($2^{3-1}+1=5$th) decides the function (for example, 111 will give different outcomes). For $y=1$, $f(0)=f(1)=1$, and/or $f(0)=1, f(1)=0$ we get an equivalent result.

Quantum mechanics, on the other hand, gives the answer with just one call of the function. To see this, let us apply the Hadamard transform to the f-CNOT outcome given by Eq. (3.68). This equation consists of products of terms which are of the following two forms (up to signs and normalization factors)

$$|x_{\text{out}-\text{constant } f}\rangle = \underbrace{(|0\rangle+|1\rangle)(|0\rangle+|1\rangle)\cdots(|0\rangle+|1\rangle)}_{n \text{ times}}, \tag{3.69}$$

$$|x_{\text{out}-\text{balanced } f}\rangle = \underbrace{(|0\rangle-|1\rangle)(|0\rangle-|1\rangle)\cdots(|0\rangle-|1\rangle)}_{n \text{ times}}, \tag{3.70}$$

as follows from Eqs. (3.58), (3.59), and (3.62). The Hadamard transform applied to these forms yields (up to signs and normalization factors)

$$\hat{\text{H}}|x_{\text{out}-\text{constant } f}\rangle = |\underbrace{00\ldots0}_{n \text{ times}}\rangle, \tag{3.71}$$

$$\hat{\text{H}}|x_{\text{out}-\text{balanced } f}\rangle = |\underbrace{11\ldots1}_{n \text{ times}}\rangle, \tag{3.72}$$

as shown in Eqs. (3.60) and (3.61).

Constant functions f give the outputs (3.69) and (3.71) and therefore we obtain the following measurement result: $|\langle 11\ldots1|00\ldots0\rangle|^2=0$ (cf. Fig. 3.20, p. 175)). Balanced functions f give the outputs (3.70) and (3.72), and $|\langle 11\ldots1|11\ldots1\rangle|^2=1$. In effect, this means that the Deutsch-Jozsa algorithm requires n steps to distinguish constant from balanced functions with *certainty*, as opposed to any classical algorithm that requires exponentially more steps ($2^{n-1}+1$). However, we should note here that we obtain Eqs. (3.69) and (3.70) only after calling the function f to act on all the qubits simultaneously and therefore on superposition given by Eq. (3.68).

Therefore, although $|x_{\text{out}}\rangle$ factors into a product of states of n qubits, we have to carry out measurements on all n qubits to obtain our result with certainty. More precisely, exponential separation between the classical and quantum efficiencies requires an absolute precision of measurement—we must have $\epsilon = 0$, where ϵ is the error in the result. In case we know that $\epsilon = 0$, we actually have to measure only one qubit to determine the result. If $\epsilon > 0$ the final result is uncertain because then the superposition (3.68) does not even partly factor into a product (Eqs. (3.69) and (3.70)). This means that if there were a bit-flip with one qubit, the final states of others would be affected also, and we would *not* obtain a product of correct terms for unaffected qubits and a "wrong" term for the affected qubit. Also, with $\epsilon > 0$, the complexity of classical measurement does not grow exponentially beyond a certain point. To see this, it suffices to pick K randomly chosen inputs and evaluate our function f on them. For K such that $2^{-K} < \epsilon$ we get the same answer independent of the total number of qubits, with a probability of error less then 2^{-K} [Jozsa, 2000].

A variation of the Deutsch-Jozsa algorithm that exhibits an exponential separation from its classical counterpart even when there is a measurement error given by Bernstein and Vazirani [Bernstein and Vazirani, 1997; Cleve et al., 1998; Jozsa, 2000]. The *Bernstein-Vazirani algorithm* employs the function $f\colon \{0,1\}^n \to \{0,1\}$ defined as

$$f(x) = (a_1 \wedge x_1) \oplus \cdots \oplus (a_1 \wedge x_1) \oplus b = (a \cdot x) \oplus b, \tag{3.73}$$

where $a \in \{0,1\}^n$ and $b \in \{0,1\}$. We can determine whether f is constant or balanced as follows

- The function f given by Eq. (3.73) is constant if $a = 00\ldots0$.
- The function f given by Eq. (3.73) is balanced if $a \neq 00\ldots0$.

The circuit diagram shown in Fig. 3.21 applies.

Classical determination of a requires n f-CNOT operations, because a consists of n bits and we have to use n different x's to check them all. Quantum mechanics requires just one call to f because Eq. (3.68) becomes [Cleve et al., 1998]

$$\frac{1}{\sqrt{2^{n+1}}}\sum_{j=0}^{2^n-1}(-1)^{(a\cdot j)\oplus b}|j\rangle(|0\rangle - |1\rangle) \tag{3.74}$$

and the Hadamard transform applied to Eq. (3.74) yields

$$\frac{1}{\sqrt{2^{n+1}}}\sum_{j=0}^{2^n-1}\sum_{k=0}^{2^n-1}(-1)^{j\cdot(a\oplus k)}|k\rangle(|0\rangle - |1\rangle) = (-1)^b|a\rangle(|0\rangle - |1\rangle). \tag{3.75}$$

A measurement analogous to the one carried out for the Deutch-Jozsa algorithm then gives the value of a.

The Bernstein-Vazirani algorithm requires polynomial time on a quantum computer as opposed to super-polynomial[12] on a classical computer even with an error $\epsilon > 0$ present. This result was improved by Daniel Simon [Simon, 1997], whose algorithm (called *Simon's algorithm*) requires an exponential time on a classical computer and again only polynomial time on a quantum computer. Simon's algorithm led Peter Shor to formulate his celebrated algorithm [Shor, 1997].

3.3.3 Shor's Algorithm

On pp. 33-34 we presented a kind of "physical calculation" by means of interference on a Mach–Zehnder interferometer. By using chosen phases with particular periods, we were able to factor (comparatively small) numbers in a polynomial time. The idea could be seen as a special case of Peter Shor's quantum algorithm [Shor, 1994; Shor, 1997] for factoring arbitrary whole numbers in particular, because the latter algorithm is also based on finding periods of the functions it uses.

Classically, we can relate the problem of factoring numbers and finding the periods of related functions in the following way. Let us factor the number $N = 15$, in particular because it is the first number that has been factored on a real (NMR) quantum computer in 2001 [Vandersypen et al., 2001].

We choose another integer a that is relatively prime (coprime) (see p. 67) to N, i.e., $\gcd(a, N) = 1$, where gcd is the *greatest common divisor.*[13] So, a could be any number from the set $\{2, 4, 7, 8, 11, 13, 14\}$. Let us choose $a = 13$. So, $a < N = 15$. Next, we consider the powers of a modulo N as shown in Table 3.1. For example, $13^1 \pmod{15} =$

Table 3.1. Powers of a modulo N exhibit periodicity. Here, the period (order) is $r = 4$, $a = 13$, $N = 15$, and $f(r) = a^r (\mathrm{mod}\, N)$.

r	1	2	3	4	5	6	7	8	9	10	11	12	13	14
$f(r)$	13	4	7	1	13	4	7	1	13	4	7	1	13	4

[12] For example, the complexity of the general number field sieve time algorithm (see p. 35) is at the same time subexponential and super-polynomial.

[13] If $\gcd(a, N) > 1$, we do not need a quantum computer for the job because then Euclid's algorithm gives a way of calculating the gcd of two numbers, without listing their divisors. It runs as follows: N/a gives a remainder of o, a/o gives a remainder of p, o/p gives a remainder of r,... u/v gives a remainder of w, and v/w gives no remainder, so $\gcd(a, N) = w$. (It is easier and faster to divide numbers then to factor them.) Euclid's algorithm requires polynomial time.

remainder of $13/15 = 13$, $13^2 \pmod{15}$ = remainder of $169/15 = 4$, etc., and we get that the period of $a^r \pmod N$ is 4. At the same time, the period is the smallest power for which we get $f(r) = 1$, i.e., for which we have

$$a^r = 1 \pmod N, \qquad r \text{ smallest.} \tag{3.76}$$

This equality hinges on the fact that a is relatively prime to N—otherwise we cannot get a remainder of 1. Such smallest r is called the *order* of a modulo N. Now, Eq. (3.76) yields

$$(a^{r/2} - 1)(a^{r/2} + 1) = 0 \pmod N = kN, \tag{3.77}$$

where k is a nonnegative integer, provided the smallest such r is even as in our case ($r = 4$). In the above example we have

$$(13^2 - 1)(13^2 + 1) = 168 \times 170 = 28560 = 1904 \times 15. \tag{3.78}$$

This means that ($\gcd(168, 15) > 1$, $\gcd(170, 15) > 1$), and we can make use of Euclid's algorithm.

In Eq. (3.77), when neither $a^{r/2} - 1$ nor $a^{r/2} + 1$ is a multiple of N, then—since their product is a multiple of N—the greatest common denominators of $a^{r/2} - 1$ and N and of $a^{r/2} + 1$ and N yield factors of N in polynomial time by means of Euclid's algorithm. In our example, $\gcd(168, 15) = 3$ and $\gcd(170, 15) = 5$.

It can be shown that the probability that r is even[14] and that $a^{r/2} \pm 1$ are not exact multiples of N (as we assumed above) is always $\geq 1/2$ [Ekert and Jozsa, 1996]. After R repetitions of the procedure, the probability of success in factoring N will be $\geq 1 - 2^{-R}$.

Thus, we have arrived at a classical procedure in which the only part which requires exponential time is determining the smallest period of $f(r)$, i.e., the order of a modulo N. This is the point where Shor's quantum procedure—which can treat all r's at once—starts.

We begin with two registers of qubits. The first register, $|x\rangle$, contains phase variables $x \in \{0, 1, 2, \dots\}$ and the second one, $|f(x)\rangle$, the values of the periodic function $f(x + r) = f(x)$, where $r \in \{0, 1, 2, \dots\}$ is the unknown least period we want to find.

The initial values of the first and second registers are

$$|\Psi_1\rangle = |0\rangle|0\rangle = \overbrace{\underbrace{|0\rangle|0\rangle \dots |0\rangle}_{n \text{ times}}}^{\text{first register}} \overbrace{\underbrace{|0\rangle|0\rangle \dots |0\rangle}_{m \text{ times}}}^{\text{second register}}, \tag{3.79}$$

[14] For odd r's the procedure is not essentially different (although a little bit more involved) [Pittenger, 1999].

where we assume that we have enough qubits to carry out the algorithm, i.e., that $N \leq 2^n < 2N^2$ (otherwise the period r might not divide 2^n) and that m is also big enough to provide us with a sufficient number of states in the second register. We create the superposition by applying a Hadamard gate to each qubit from the first register as we did in Eq. (3.67) and as shown in Fig. 3.22:

$$\begin{aligned} \hat{\mathrm{H}}^{(n)}|00\ldots0\rangle &= \hat{\mathrm{H}}^{(n)}|0\rangle|0\rangle\ldots|0\rangle = \hat{\mathrm{H}}|0\rangle\hat{\mathrm{H}}|0\rangle\ldots\hat{\mathrm{H}}|0\rangle \\ &= \frac{1}{\sqrt{2^n}}(|0\rangle+|1\rangle)(|0\rangle+|1\rangle)\ldots(|0\rangle+|1\rangle) \\ &= \frac{1}{\sqrt{2^n}}\sum_{j=0}^{2^n-1}|j\rangle, \end{aligned} \tag{3.80}$$

where $|j\rangle$, $j = 0, \ldots, 2^{n-1}$ are defined as

$$|0\rangle = \underbrace{|0\rangle\ldots|0\rangle}_{n}, \quad |1\rangle = \underbrace{|0\rangle\ldots|1\rangle}_{n}, \quad \ldots \quad |2^{n-1}\rangle = \underbrace{|1\rangle\ldots|1\rangle}_{n}. \tag{3.81}$$

The total function at this stage of our procedure reads

$$|\Psi_2\rangle = \frac{1}{\sqrt{2^n}}\sum_{j=0}^{2^n-1}|j\rangle|0\rangle, \tag{3.82}$$

where $|0\rangle$ in the second register means $|0\rangle|0\rangle\ldots|0\rangle$.

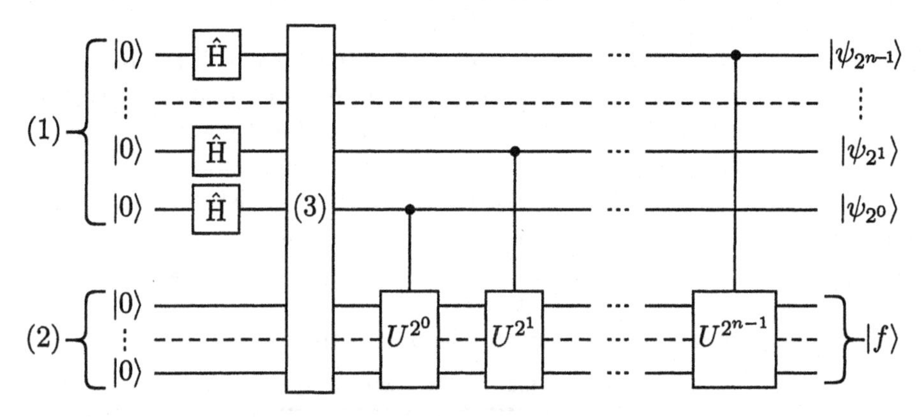

Figure 3.22. Period-finding algorithm circuit diagram. (1) First register, n qubits. (2) Second register, m qubits. Box (3) represents the computation of $f(j)$ (j's are from the 1st register) carried out in a 3rd register, with the obtained values stored in the 2nd register; values of the 1st register remain unchanged. Operators $U = e^{i\phi}$ implement the quantum Fourier transform [Eq. (3.86)]—the product of the relevant $|\psi\rangle$ states given by Eq. (3.88) is equal to it, as follows from Eq. (3.89).

In our example above ($N = 15$, $a = 13$), we can choose $n = 4$ qubits ($2^4 = 16 > 15$) and from Eq. (3.82) we get

$$\begin{aligned}|\Psi_2\rangle &= \frac{1}{\sqrt{2^4}}(|0\rangle + |1\rangle + \cdots + |2^4 - 1\rangle)|0\rangle \\ &= \frac{1}{\sqrt{2^4}}(|0000\rangle|0...0\rangle + |0001\rangle|0...0\rangle + \cdots + |1111\rangle|0...0\rangle). \quad (3.83)\end{aligned}$$

Next, in a separate, third register, we compute $f(j) = a^j$ (mod$N$) for each j from the first register and store their values in the second register. Thus we obtain

$$|\Psi_3\rangle = \frac{1}{\sqrt{2^n}} \sum_{j=0}^{2^n-1} |j\rangle|f(j)\rangle. \quad (3.84)$$

This stage of the procedure is denoted by box (3) in Fig. 3.22. In our case, $f(j)$ is given by the second row of Table 3.1 (except for $f(0) = 13^0$ (mod 15) $= 1$), and we have

$$\begin{aligned}|\Psi_3\rangle &= \frac{1}{\sqrt{2^4}}(|0\rangle|1\rangle + |1\rangle|13\rangle + |2\rangle|4\rangle + |3\rangle|7\rangle + |4\rangle|1\rangle + |5\rangle|13\rangle \cdots |15\rangle|7\rangle) \\ &= \frac{1}{\sqrt{2^4}}(|0000\rangle|0001\rangle + |0001\rangle|1101\rangle + |0010\rangle|0100\rangle + |0011\rangle|0111\rangle \\ &\quad + |0100\rangle|0001\rangle + |0101\rangle|1101\rangle \cdots + |1111\rangle|0111\rangle). \quad (3.85)\end{aligned}$$

Here, we have chosen $m = n = 4$. The number of qubits in the second register, m, can be smaller, but the number of states, M has to be big enough to enable a reliable measurement of the period. For example, in the experiment mentioned above, three qubits ($m = 3$) were put in the second register [Vandersypen et al., 2001].

To find the period r of our function f, we apply a *quantum Fourier transform* on each state from the first register, because it turns out that for phases inversely proportional to the period (r) we have constructive interference and for all the other phases destructive interference. The quantum Fourier transform on an orthonormal basis $|0\rangle, \ldots, |N - 1\rangle$ is defined as the following linear operator:

$$|j\rangle \longrightarrow \frac{1}{\sqrt{K}} \sum_{k=0}^{K-1} e^{2\pi ijk/K}|k\rangle. \quad (3.86)$$

Thus the function describing the fourth stage of our procedure reads

$$|\Psi_4\rangle = \frac{1}{2^n} \sum_{k=0}^{2^n-1} \sum_{j=0}^{2^n-1} e^{2\pi ijk/2^n}|k\rangle|f(j)\rangle. \quad (3.87)$$

For each k, the corresponding terms of this function can be implemented by the quantum circuit shown in Fig. 3.22, where

$$|\psi_{2^0}\rangle = \frac{1}{\sqrt{2}}(|0\rangle + e^{2^{1-n}\pi ik}|1\rangle), \quad |\psi_{2^1}\rangle = \frac{1}{\sqrt{2}}(|0\rangle + e^{2^{2-n}\pi ik}|1\rangle), \ \ldots$$
$$|\psi_{2^{n-2}}\rangle = \frac{1}{\sqrt{2}}(|0\rangle + e^{2^1\pi ik}|1\rangle), \quad |\psi_{2^{n-1}}\rangle = \frac{1}{\sqrt{2}}(|0\rangle + e^{2^0\pi ik}|1\rangle), \quad (3.88)$$

as follows from

$$(|0\rangle + e^{2^{1-n}\pi ik}|1\rangle)(|0\rangle + e^{2^{2-n}\pi ik}|1\rangle)$$
$$\cdots(|0\rangle + e^{2^1\pi ik}|1\rangle)(|0\rangle + e^{2^0\pi ik}|1\rangle) = \sum_{j=0}^{2^n-1} e^{2\pi ijk/2^n}|j\rangle. \quad (3.89)$$

Now, a measurement of the second register selects all those $|f(j)\rangle$ that are equal to each other. It singles out the j's of the corresponding states $|j\rangle$ given by Eq. (3.89). For example, in Eq. (3.85) a detection of state $|13\rangle$ from the second register singles out $|1\rangle$, $|5\rangle$, $|9\rangle$, and $|13\rangle$ from the first register, i.e., it selects $j = 1, 5, 9, 13$. The probability that a measurement of the first register will confirm that its qubits are in one of the Fourier states, $|k\rangle$, follows from Eq. (3.89):

$$\Pr(k) = \frac{1}{2^{2n}}\left|\sum_{j=0;\,\mathrm{sel.}j\text{'s}}^{2^n-1} e^{2\pi ijk/2^n}\right|^2, \qquad \text{for } f(j)\text{-selected } j\text{'s}. \quad (3.90)$$

Let us estimate this probability, following Shor's reasoning [Shor, 1997]. The j's are either integer multiple of the period r (see Eq. (3.76)), lr, or, due to periodicity (see Table 3.1), shifted integer multiples $lr+q$, where l and q are nonnegative integers. In our case (see Eq. (3.85)), for instance, for $f(j) = 1$ we have $j = 0, 4, 8, 12$, i.e., $j = 4l$, $l = 0, 1, 2, 3$, and for $f(j) = 13$ we have $j = 1, 5, 9, 13$, i.e., $j = 4l + q$, $l = 0, 1, 2, 3$, $q = 1$. So, let us see what we would get if we knew the period r in advance and if we assumed that $q \approx 0$ (in comparison with 2^n). Then r would divide 2^n exactly and $l = j/r$ in the sum of Eq. (3.90) would range from $l = 0$ to $l = 2^n/r - 1$. The normalization factor in Eq. (3.84) becomes $\sqrt{r/2^n}$. Thus we can write Eq. (3.90) in the following form:

$$\Pr(k) = \frac{r}{2^{2n}}\left|\sum_{l=0}^{\frac{2^n}{r}-1} e^{2\pi i(lr+q)k/2^n}\right|^2 = \frac{r}{2^{2n}}\left|\sum_{l=0}^{\frac{2^n}{r}-1} e^{2\pi ilrk/2^n}\right|^2. \quad (3.91)$$

If $k = 2^n p/r$, where p is a nonnegative integer, then $\Pr(k) = 1/r$ because $e^{2\pi ilp} = 1$ for $l, p = 0, 1, 2, \ldots$. If not, the terms in Eq. (3.91) cancel each other. Hence, we obtain

$$\Pr(k) = \begin{cases} \frac{1}{r} & \text{if } k \text{ is a multiple of } \frac{2^n}{r} \\ 0 & \text{otherwise} \end{cases} \tag{3.92}$$

The state of the qubits from the first register for a chosen k at this final stage of our procedure is

$$\begin{aligned} |\Psi_5\rangle &= \sqrt{\frac{r}{2^{2n}}} \sum_{l=0}^{\frac{2^n}{r}-1} e^{2\pi i(lr+q)k/2^n}|k\rangle \\ &= \sqrt{\frac{r}{2^{2n}}} \left(\sum_{l=0}^{\frac{2^n}{r}-1} e^{2\pi i lrk/2^n} \right) e^{2\pi i qk}|k\rangle. \end{aligned} \tag{3.93}$$

For $k = 2^n p/r$ the expression in the brackets on the right hand side of Eq. (3.93) is equal to $2^n/r$ and the corresponding state is

$$|\Psi_5\rangle = \frac{1}{\sqrt{r}} e^{2\pi i qp/r}|p\frac{2^n}{r}\rangle. \tag{3.94}$$

By measuring the first register, we detect the state with probability $1/r$ [Cleve et al., 1998]. The procedure therefore requires repeated measurements, but the number of the repetitions required does not grow exponentially with the number of qubits. We obtain $2^n p/r$, $p = 0, 1, 2, \ldots$ where the values of p are equiprobable. Then we extract r from the obtained $2^n p/r$ by using the classical *continued fraction expansion* (polynomially complex procedure) on a classical computer.

In case when r does not divide 2^n we arrive at essentially the same result. The only difference is that the coefficients of the Fourier transform are no longer represented by sharp values (see Eq. (3.92)) but by peaks on the closest integers to the multiples of $2^n/r$ [Shor, 1997].

To see that the above quantum factoring algorithm is of a polynomial complexity, it is sufficient to sum up the steps of the quantum Fourier transform procedure, altogether about n^2 steps—each superposition requires only one step. By contrast, the classical Fourier transform takes about $n2^n$ steps, thus requiring an exponentially growing time. By examining the period-finding procedure, we can understand the difference as follows. If we tried to find the period of function $|\Psi_4\rangle$ in Eq. (3.84) directly, it would require a classical trial-and-error procedure and therefore an exponential time. Only after we apply the Fourier transform can we obtain the function $|\Psi_5\rangle$ (Eqs. (3.93) and (3.94)), which enables detection of the period in a polynomial number of steps and a polynomial time.

3.3.4 Quantum Simulators

Practically all known quantum algorithms and the transforms they use are based on Fourier transforms. The Hadamard transform

$$\hat{\mathrm{H}}|0\rangle = \frac{1}{\sqrt{2}}(|0\rangle + |1\rangle), \qquad \hat{\mathrm{H}}|1\rangle = \frac{1}{\sqrt{2}}(|0\rangle - |1\rangle). \tag{3.95}$$

as well as the f-CNOT (Def. 3.3, p. 174, Eq. (3.62))

$$\begin{aligned} f\text{-}\widehat{\mathrm{CNOT}}\ & [(|0\rangle + |1\rangle)(|0\rangle - |1\rangle)] \\ & = [(-1)^{f(0)}|0\rangle + (-1)^{f(1)}|1\rangle](|0\rangle - |1\rangle) \end{aligned} \tag{3.96}$$

are just special cases of Fourier transforms.

Deutsch's, the Deutsch-Jozsa, the Bernstein-Vazirani, Simon's, and Shor's algorithms are all based on Fourier transforms, as well as *Grover's quantum search algorithm.*[15]

Another common property of these algorithms is that they all reduce to finding eigenvalues and eigenvectors of unitary operators. It turns out that the latter problem can be solved by means of quantum Fourier transforms. For instance, the phase factor of $(-1)^{f(x)}$ from Eqs. (3.62) and (3.96) is the eigenvalue of the state of the bottom control qubit in Figs. 3.20 (p. 175) and 3.21 (p. 177) under the action of an operator defined by mapping $|y\rangle \rightarrow |y \oplus f(x)\rangle$. For Deutsch's algorithm the mapping is determined by Eqs. (3.57)–(3.62). On the other hand, Shor's algorithm can be viewed as a procedure for finding the eigenvalues that correspond to an eigenstate of an operator that maps $|x\rangle$ to $|a^x \pmod N\rangle$ (see Eq. (3.82)) [Cleve et al., 1998].

Quantum algorithms for finding eigenvalues and eigenvectors of unitary operators can actually be treated as a general problem and shown to provide an exponential speed increase compared to the standard way of solving such problems on a classical computer. Here, we start with the approach put forward by Abrams and Lloyd [Abrams and Lloyd, 1999] and Zalka [Zalka, 1998]. We consider the time-evolution unitary operator

$$\hat{U} = e^{-i\hat{H}t/\hbar}, \tag{3.97}$$

where the operator H is the Hamiltonian of a system we consider. Abrams and Lloyd considered local systems such as van der Waals gases, Heisenberg spin systems, and strong and weak interactions, while Zalka's approach is more general but neglects the imperfections of realistic quantum computers.

[15]Grover's algorithm offers a quadratic speedup with respect to classical search algorithms.

Let us denote the eigenvector and eigenvalue of $\hat{U}$ by V and λ_ν, respectively. In general, with a quantum computer composed of qubits—two-level systems—we can only approximate the true eigenvector V. We denote an approximate eigenvector of $\hat{U}$ by V_a and the approximate eigenvalue by λ_a and assume that $|\langle V_a|V\rangle|^2$ is not exponentially small. We then find λ_ν in a time inversely proportional to $\epsilon|\langle V_a|V\rangle|^2$ with a polynomially decreasing accuracy ϵ. By contrast, all known classical algorithms do the job with exponentially decreasing accuracy.

We use three registers. The first register contains n *index qubits*, the second register contains m *Fourier qubits* that determine the space in which $\hat{U}$ acts, and the third one contains w *storage qubits*. The initial state is

$$|\Psi_0\rangle = |0\rangle|V_a\rangle, \tag{3.98}$$

where, for example, $|0\rangle$ is defined as in Eq. (3.79).

The superposition of the qubit states from the first register can be created as in Eqs. (3.80) and (3.81) by applying Hadamard gates to each qubit:

$$|\Psi_1\rangle = \frac{1}{\sqrt{2^n}}\sum_{j=0}^{2^n-1}|j\rangle|V_a\rangle. \tag{3.99}$$

Then in the third register, we compute $\hat{U}^j|V_a\rangle$ (taking into account that $\hat{U}|V_a\rangle = \lambda_a|V_a\rangle$) for each j from the first register and store their values in the second register. By writing λ_a as $\exp(i\omega_a)$, we obtain

$$|\Psi_2\rangle = \frac{1}{\sqrt{2^n}}\sum_{j=0}^{2^n-1}|j\rangle\hat{U}^j|V_a\rangle = \frac{|V_a\rangle}{\sqrt{2^n}}\sum_{j=0}^{2^n-1}e^{i\omega_a j}|j\rangle. \tag{3.100}$$

If we now perform the quantum Fourier transform as in Eq. (3.87), we will detect the phase ω_a and thereby the eigenvalue λ_a, which will approximate the true eigenvalue λ_ν. To do any concrete computation of eigenvalues and eigenvectors of the operator $\hat{U}$ one has specify the Hamiltonian $\hat{H}$. We start with the one corresponding to N mutually interacting particles (V_{ij}, $i,j = 1,\ldots,N$) in an external potential (V_i, $i = 1,\ldots,N$), i.e., with a corresponding Schrödinger equation. To find solutions to the Schrödinger equation means to find energy eigenstates and eigenvectors for the following Hamiltonian:

$$\hat{H} = \sum_{i=1}^{N}(\hat{T}_i + V_i) + \sum_{i>j}^{N} V_{ij}, \tag{3.101}$$

where T_i is the kinetic energy operator.

Let us consider the simplest possible Schrödinger equation—of a single quantum particle in one dimension—to see how it can be simulated on a quantum computer:

$$i\frac{\partial\Psi(x)}{\partial t} = \hat{H}\Psi(x) = \left[\frac{\hat{P}^2}{2m} + V(\hat{X})\right]\Psi(x), \tag{3.102}$$

where we have set $\hbar = 1$. Our final goal is to simulate a chosen quantum system—in effect its Schrödinger equation—by qubits in a quantum computer.

The first thing we have to do with a Schrödinger equation to solve it, i.e., simulate it, by means of qubits is to discretize it, because the "standard quantum computer" that uses qubits (two-state systems) cannot give a continuous function as its output. A continuous function that is the exact solution to the Schrödinger equation (3.102) can only be approximated by discrete measurement results that we obtain at the end of a calculation. It is similar to the procedure of reading off the phase on p. 185. We carry out the discretization by imposing periodicity on the coefficients of the following expansion of state Ψ:

$$|\Psi\rangle = \sum_{j=0}^{2^n-1} a_j|j\rangle, \tag{3.103}$$

where $|j\rangle$ is the basis state corresponding to the binary representation of the number i (see Eq. (3.82)).

The time evolution can be implemented through small Δt steps by means of the time evolution operator [Zalka, 1998]:

$$\hat{U}(\Delta t) = e^{-i\Delta t\hat{H}} = e^{-i\Delta t[1/2m\hat{P}^2+V(\hat{X})]}. \tag{3.104}$$

For noncommuting variables x and y (as in Eq. (3.104)), we have $e^{x+y} \neq e^x e^y$ and must use the Baker-Campbell-Hausdorff series [Reinsch, 2000]

$$\exp(x)\exp(y) = \exp\left[\sum_{j=1}^{\infty} f_j(x,y)\right]. \tag{3.105}$$

The first few terms of the series are

$$\begin{aligned}
f_1(x,y) &= x + y,\\
f_2(x,y) &= \frac{1}{2}(xy - yx) = \frac{1}{2}[x,y],\\
f_3(x,y) &= \frac{1}{12}(x^2y + xy^2 - 2xyx + y^2x + yx^2 - 2yxy).
\end{aligned} \tag{3.106}$$

Thus in the first order approximation we obtain

$$\hat{U}(\Delta t) \approx e^{-i\Delta t\hat{H}\hat{P}^2/2m}e^{-i\Delta tV(\hat{X})}. \tag{3.107}$$

To implement it in a quantum computer we go to momentum-space by a standard procedure, using the following Fourier transform of the wave function Ψ [Messiah, 1965]:

$$\Phi(p) = \frac{1}{\sqrt{2\pi}}\sum_x \Psi(x)e^{-ipx}. \tag{3.108}$$

Function $\Phi(p)$ can be interpreted as a linear combination of elementary waves $\exp(ipx)$ of a well defined position, each elementary wave having a coefficient $\Psi(x)/\sqrt{2\pi}$. Then we can apply the quantum Fourier transform algorithm from Sec. 3.3.3.

The unitary transformations we obtain in Eq. (3.107), which are of the type

$$|j\rangle \rightarrow e^{icF(j)}|j\rangle, \tag{3.109}$$

can be carried out with the following steps [Zalka, 1998] analogous to those in Sec. 3.3.3:

$$|j,0\rangle \rightarrow |j,F(j)\rangle, \quad |F(j)\rangle \rightarrow e^{icF(j)}|F(j)\rangle, \quad |j,F(j)\rangle \rightarrow |j,0\rangle. \tag{3.110}$$

The first and last steps are, in effect, n quantum calculations carried out in parallel. The second step requires about n calculational steps. This, together with the Fourier transform above, gives us altogether n^2 computational steps. A classical calculation of the same problem requires at least 2^{2n} computational steps, because just to write down a wave function for n particles requires 2^n memory sites and to compute its time evolution requires the exponentiation of $2^n \times 2^n$ matrices [Lloyd, 1996].

To achieve higher precision, we have to use higher terms from Eq. (3.106). For example, in the second order approximation, we get

$$e^{t(\hat{A}+\hat{B})} \approx e^{t\hat{A}}e^{t\hat{B}}e^{-t^2[\hat{A},\hat{B}]/2} \tag{3.111}$$

from which we derive

$$e^{t(\hat{A}+\hat{B})} \approx e^{t\hat{A}/2}e^{t\hat{B}}e^{t\hat{A}/2}. \tag{3.112}$$

To generalize the simulation to N particles in three dimensions with the potential $V(\hat{\boldsymbol{X}}_1, \hat{\boldsymbol{X}}_2, \dots)$, we use $3N$ quantum registers and apply unitary transformations on several registers, e.g.

$$|j,j',j''\rangle \rightarrow e^{icF(j,j',j'')}|j,j',j''\rangle. \tag{3.113}$$

In the case of more general Hamiltonians, for instance excited states of molecules, we no longer have an obvious and straightforward algorithm, although simulations of universal local Hamiltonians describing close range interactions (strong and weak interactions, van der Waals gases, etc.) are feasible [Lloyd, 1996; Abrams and Lloyd, 1997; Abrams and Lloyd, 1999; Gramss, 1998; Boghosian and Taylor IV, 1998]. No algorithms are known that implement Fourier transforms for simulating and determining the evolution of general arbitrary systems.

What quantum computers lack is a general quantum algebra that would correspond to the Boolean algebra on classical computers.

3.4 Quantum Turing Machines vs. Quantum Algebra

The lack of a general quantum algebra, at first glance, seems to clash with the fact that there is a well-defined concept of the *quantum Turing machine*. To see why a quantum Turing machine does not require a well defined general quantum algebra, we will consider its definition [Bernstein and Vazirani, 1997; Gruska, 1999], starting with a definition of the *probabilistic Turing machine* [Gruska, 1999, Sec. 1.3.1].

DEFINITION 3.4 *A* probabilistic Turing machine *is a triple* $(\mathcal{U}, \mathcal{S}, \delta_{\rm p})$, *where* $\delta_{\rm p}$, *the* probabilistic transition function, *is a function*

$$\delta_{\rm p} : \mathcal{U} \times \mathcal{S} \longrightarrow \widetilde{[0,1]}^{\,\mathcal{U}\times\mathcal{S}\times\{l,r,h\}}, \tag{3.114}$$

where $\mathcal{U}$ is a finite alphabet as defined on p. 2, $\mathcal{S}$ is a finite set of states (see p. 2), {l,r,h} is the set of actions (see p. 3), and $\widetilde{[0,1]}$ is the set consisting of $p_j \in [0,1]$, $j = 1,\ldots,M$ (M is the number of output states) such that there is an algorithm that computes p's.

The probabilistic transition function $\delta_{\rm p}$ assigns probabilities p_j, $j = 1,\ldots,M$ to each output state c_j (see Table 1.1, p. 3) so as to satisfy

$$\sum_{j=1}^{M} p_j = 1. \tag{3.115}$$

Each p_j is the probability of detecting the output state c_j.

We can make use of the above definition of a probabilistic Turing machine and its properties to define a quantum Turing machine in the following way.

DEFINITION 3.5 *A* quantum Turing machine *is a triplet* $(\mathcal{U}, \mathcal{S}, \delta_{\rm q})$, *where* $\delta_{\rm q}$, *the* quantum transition function, *is a function*

$$\delta_{\rm q} : \mathcal{U} \times \mathcal{S} \longrightarrow \widetilde{C}_{[0,1]}^{\,\mathcal{U}\times\mathcal{S}\times\{l,r,h\}}, \tag{3.116}$$

where $\widetilde{C}_{[0,1]}$ is the set consisting of $\alpha_j \in C_{[0,1]}$, $j = 1, \ldots, M$ (where C is the field of complex numbers) such that there is an algorithm that computes the real and imaginary parts of α's to within 2^{-M} in time polynomial in M.

The quantum transition function δ_q assigns probability amplitudes α_j, $j = 1, \ldots, M$ (complex numbers the absolute values of which are in the interval $[0, 1]$) to each output state so as to satisfy

$$\sum_{j=1}^{M} |\alpha_j|^2 = 1. \tag{3.117}$$

Each $|\alpha_j|^2$ is the probability of detecting the output c_j.

Hence, at the final output level, probabilistic and quantum Turing machines behave in a similar way. However, in general—due to differences at intermediate levels—they behave very differently. Let us assume that each amplitude α_j is sum of subamplitudes: $\sum_k \beta_k$. Here we can have either destructive $(|\sum_k \beta_k|^2 < \sum_k |\beta_k|^2)$ or constructive $(|\sum_k \beta_k|^2 > \sum_k |\beta_k|^2)$ interference that we cannot have with p_j. Thus, as with Shor's algorithm (see Eq. (3.92)), we can obtain a definite probability for some input values and 0 for all the others. Another essential difference is that the transition matrices corresponding to each quantum Turing machine are unitary and that quantum Turing machines are therefore reversible.

The fact that a quantum Turing machine is defined on the field of complex numbers is not an obstacle for an implementation of its algorithms on a realistic quantum circuit. Any quantum Turing machine defined on the field of complex numbers can be simulated by another quantum Turing machine, all transitions of which have real amplitudes [Gruska, 1999, Theorem 4.2.17]. Actually this is so because (the Hilbert space) quantum mechanics can be formulated over three fields: the field of real numbers, the field of complex numbers, and the (skew) field of quaternions [Holland, JR., 1995].

Since we literally transcribe all the amplitude algorithms based on the Fourier transform for the quantum Turing machine and since actions of all quantum circuits can be viewed as amplitude handling, we can conclude that quantum circuits can compute at least everything the quantum Turing machine can compute. On the other hand, we have seen that that the universal Fourier transform for solving a universal quantum system is not known. This means that a universal quantum algebra is not needed for quantum circuits nor for quantum Turing machine. However, such a quantum algebra would enable us to implement any Schrödinger equation directly into a quantum computer, thus providing us with a universal quantum translator for Schrödinger equations. The algebra would correspond to Boolean algebra in classical computers, but there would be a crucial difference. Any classical and quantum problem alike has to be written in a binary code before we put it into the classical computer. On the other hand, any Schrödinger equation would simply be an expression in quantum algebra, and its simulation on a quantum computer would just be an equivalent expression in the same algebra. Recall that the complexity of trial division is exponential just because we have to write the numbers and instructions in a binary form (compare Fig. 1.14 (a) and 1.14 (b), p. 36). Recall also that any realistic quantum Fourier transform from the previous sections was an approximation done by discretizing the systems we wanted to solve.

There have been many attempts to find a quantum algebra—some of them being C^* algebra, von Neumann algebra, Baer *-rings, projective geometries, and Hilbert lattices—but so far none of them have turned out to be successful. In what follows, we shall attempt to show the main reason why the problem is so hard to crack. To do so, we will use the Hilbert lattice approach since it is the easiest to understand with the tools we introduced in the previous sections.

Let us start with a few definitions (cf. Def. 1.2, p. 8).

Definition 3.6 *A* lattice *is an algebra* $\mathcal{L} = \langle L, \cup, \cap \rangle$, *where the operations* $\cup$ *and* $\cap$ *are called* join *and* meet *respectively, such that the following conditions are satisfied for any* $A, B, C \in L$:

1. Commutativity (a) $A \cup B = B \cup A$, (b) $A \cap B = B \cap A$
2. Associativity (a) $(A \cup B) \cup C = A \cup (B \cup C)$, (b) $(A \cap B) \cap C = A \cap (B \cap C)$
3. Absorption (a) $A \cap (A \cup B) = A$, (b) $A \cup (A \cap B) = A$,

A lattice is a partially ordered set ($A \le B \overset{\text{def}}{=} A = A \cap B$) whose elements A,B have the least upper bound $A \cup B$ and the greatest lower bound $A \cap B$.

Definition 3.7 *An* ortholattice *is an algebra* $\mathcal{OL} = \langle OL, \cup, \cap, \bar{\ }, 0, 1 \rangle$ *such that the following conditions are satisfied for any* $A, B \in OL$:

1. *The algebra* $\langle OL, \cup, \cap \rangle$ *is a lattice*
2. $A \cup \overline{A} = 1$ & $A \cap \overline{A} = 0$
3. $A \le B \Rightarrow \overline{B} \le \overline{A}$
4. $\overline{\overline{A}} = A$

The operation "$\overline{}$" is called *orthocomplementation.*

DEFINITION 3.8 *The following three conditions*

$$\begin{aligned} & & A\cap(B\cup C)=(A\cap B)\cup(A\cap C), \quad (3.118)\\ B\le A & \Rightarrow & A\cap(B\cup C)=(A\cap B)\cup(A\cap C), \quad (3.119)\\ B\le A \ \ \& \ \ C\le\overline{A} & \Rightarrow & A\cap(B\cup C)=(A\cap B)\cup(A\cap C), \quad (3.120)\end{aligned}$$

are called the distributivity, modularity, *and* orthomodularity *conditions, respectively. They have dual forms in which* $\cup$ *and* $\cap$ *are interchanged.*

One can prove the following theorem [Maeda and Maeda, 1970; Kalmbach, 1974; Mittelstaedt, 1978; Beran, 1985].

THEOREM 3.9 *An ortholattice* $\mathcal{OL}$ *in which the conditions (3.118), (3.119), (3.120) (or their dual forms) hold, for any* $A, B, C \in OL$, *is a Boolean algebra*[16] (a distributive lattice), modular lattice, *and* orthomodular lattice, *respectively. Also, in these lattices the respective conditions hold.*

Boolean algebra (specifically, the 2-valued Boolean algebra) provides the computational basis for classical computers. Modular lattices provide a computational basis for finite dimensional Hilbert spaces and therefore for spin systems—as revealed by Birkhoff and von Neumann in 1936 [Birkhoff and von Neumann, 1936]—and consequently for today's discrete quantum computers. Modular lattices correspond to the universal quantum gate operations, which are in the quantum computation literature usually called quantum logic.[17] We should be careful with the term "logic," though. Classical logic [Hilbert and Ackermann, 1950] and quantum logic [Kalmbach, 1974; Dishkant, 1974; Dalla Chiara, 1986] in the mathematical logic literature hang on the following theorem [Pavičić, 1987b; Pavičić, 1989; Pavičić and Megill, 1999].

THEOREM 3.10 *An ortholattice* $\mathcal{OL}$ *in which the following conditions hold for any* $A, B, C \in OL$

$$A \to_i B = 1 \quad \Leftrightarrow \quad A \le B, \qquad i = 0,\dots,5, \qquad (3.121)$$

$$A \equiv_i B = 1 \quad \Leftrightarrow \quad A = B, \qquad i = 0,\dots,5, \qquad (3.122)$$

[16]Then the operations $\cup$ and $\cap$ become $+$ and $\cdot$ from Def. 1.2 (p. 8), respectively.

[17]In set and lattice theory literature a *quantum logic* often means a σ*-orthomodular partially ordered set*, i.e., a σ*-orthomodular poset*, where σ indicates closure under the formation of suprema of a countable family of mutually orthogonal subsets of the poset [Pták and Pulmannová, 1991].

where

$$
\begin{aligned}
&A \to_0 B \overset{\text{def}}{=} \overline{A} \cup B, \quad A \to_1 B \overset{\text{def}}{=} \overline{A} \cup (A \cap B), \quad A \to_2 B \overset{\text{def}}{=} \overline{B} \to_1 \overline{A},\\
&A \to_3 B \overset{\text{def}}{=} (\overline{A} \cap B) \cup (\overline{A} \cap \overline{B}) \cup (A \to_1 B), \qquad A \to_4 B \overset{\text{def}}{=} \overline{B} \to_3 \overline{A},\\
&A \to_5 B \overset{\text{def}}{=} (A \cap B) \cup (\overline{A} \cap B) \cup (\overline{A} \cap \overline{B}),\\
&A \equiv_i B \overset{\text{def}}{=} (A \to_i B) \cap (B \to_i A), \qquad i = 0, \ldots, 5, \qquad (3.123)
\end{aligned}
$$

is a Boolean algebra for $i = 0$ and an orthomodular lattice for $i = 1, \ldots, 5$, and vice versa—in a Boolean algebra and in an orthomodular lattice these conditions hold.[18]

Now, a proper classical logic is, in effect, given by a system in which all the axioms of a Boolean algebra (e.g., from Def. 1.2, p. 8 or from Defs. 3.6, 3.7, and 3.8, Eq. (3.118)) are written so as to substitute $A \equiv_0 B = 1$ for any equality $A = B$ and $A \to_0 B = 1$ for any inequality $A \le B$. The same is the case with quantum (modular and orthomodular) logics (Defs. 3.6, 3.7, and 3.8, Eq. (3.119) and Eq. (3.120)) when we substitute $A \equiv_i B = 1$ for any equality $A = B$ and $A \to_i B = 1$ for any inequality $A \le B$, $i = 1, \ldots, 5$.

It is only when we demand that the conditions (3.121) and (3.122) be met that we necessarily have a Boolean algebra and an orthomodular (or modular) lattice as models for classical and quantum logics, respectively. But there are lattice models for these logics in which conditions (3.121) and (3.122) do not hold. For example, one can prove the completeness and soundness of classical logic for a lattice model that is not distributive and of quantum logic for a lattice model that is not (ortho)modular [Pavičić and Megill, 1999]. Hence, if we are only interested in obtaining working models for classical or quantum computers, we had better stay with algebras.

As we already emphasized, we do have a modular lattice model which is an algebra for quantum circuits. What we need is a more general algebra, of which the modular algebra is a special case, to include continuous variables. We do have the orthomodular algebra (lattice) which underlies any infinite-dimensional Hilbert space. Up to now, an infinite-dimensional Hilbert space has been our only tool to handle continuous variables. A considerable theoretical and experimental effort has recently been done towards developing continuous-variable quantum computers and their possible algorithms [Lloyd and Braunstein, 1999; Braunstein

[18] In any Boolean algebra (i.e., when the distributivity (3.118) is added to an orthomodular lattice), $A \to_i B$ and $A \equiv_i B$, $i = 1, \ldots, 5$ reduce to $A \to_0 B$ and $A \equiv_0 B$, respectively.

and Pati, 2003]). However, developing a discrete algebra that could generate continuous-variable Schrödinger equation results (eigenvalues and eigenstates) might have an appeal of its own, not only for running the equation on a discrete quantum computer, but also for quantizing our equations. Continuous-variable equations are primarily needed not to describe the free motion of a quantum system, as we did in Sec. 3.3.4 (p. 186), but instead to describe continuous Coulomb potentials in order to obtain position distributions of electrons, nucleons, and atoms in molecules and matter. As an example, consider a simple hydrogen atom. For the principal quantum number $n = 3$ and the orbital quantum number $l = 0$, its radial function has three maxima of which the third is the greatest, and we say that the electron is in the third shell. If the right algebra could be found, perhaps we might arrive at equations which would just give us three peaks at these maxima or a single peak at the maximum of the probability function, thus avoiding having to solve the huge matrices that we obtain by discretizing continuous functions. So, let us see how far we have advanced on the road towards algebraization of continuous Hilbert spaces.

The main idea behind representing Hilbert space by an orthomodular lattice is to add additional strengthening axioms which are still weak enough so as not to make it modular. These axioms will give us the so-called *Hilbert lattices* [Beltrametti and Cassinelli, 1981; Kalmbach, 1986].

DEFINITION 3.11 *An orthomodular lattice which satisfies the following conditions is a* Hilbert lattice, $\mathcal{HL}$.

1. Completeness: *The meet and join of any subset of an $\mathcal{HL}$ exist.*

2. Atomic: *Every non-zero element in an $\mathcal{HL}$ is greater than or equal to an atom. (An atom A is a non-zero lattice element with $0 < B \le A$ only if $B = A$.)*

3. Superposition Principle: *(The atom C is a superposition of the atoms A and B if $C \ne A$, $C \ne B$, and $C \le A \cup B$.)*

   (a) *Given two different atoms A and B, there is at least one other atom C, $C \ne A$ and $C \ne B$, that is a superposition of A and B.*

   (b) *If the atom C is a superposition of distinct atoms A and B, then atom A is a superposition of atoms B and C.*

4. Minimal length: *The lattice contains at least three elements A, B, C satisfying: $0 < A < B < C < 1$.*

One can prove the following theorem [MacLaren, 1964; Mackey, 1963; Varadarajan, 1970].

THEOREM 3.12 *For every Hilbert lattice $\mathcal{HL}$ there exists a field $\mathcal{K}$ and a Hilbert space $\mathcal{H}$ over $\mathcal{K}$ such that the set of closed subspaces of the Hilbert space, $\mathcal{C}(\mathcal{H})$ is ortho-isomorphic to $\mathcal{HL}$.*

Conversely, let $\mathcal{H}$ be an infinite-dimensional Hilbert space over a field $\mathcal{K}$ and let

$$\mathcal{C}(\mathcal{H}) \stackrel{\text{def}}{=} \{\mathcal{X} \subseteq \mathcal{H} \mid \mathcal{X}^{\perp\perp} = \mathcal{X}\} \tag{3.124}$$

be the set of all biorthogonal closed subspaces of $\mathcal{H}$. Then $\mathcal{C}(\mathcal{H})$ is a Hilbert lattice relative to:

$$A \cap B = \mathcal{X}_A \cap \mathcal{X}_B \qquad \text{and} \qquad A \cup B = (\mathcal{X}_A + \mathcal{X}_B)^{\perp\perp}. \tag{3.125}$$

In order to determine the field over which the Hilbert space in Theorem 3.12 is defined, we make use of the following theorem proved by Maria Pia Solèr [Solèr, 1995; Holland, JR., 1995].

THEOREM 3.13 *The Hilbert space $\mathcal{H}$ from Theorem 3.12 is an infinite-dimensional Hilbert space defined over a real, complex, or quaternion (skew) field if the following conditions are met:*

- Infinite orthogonality: *Any $\mathcal{HL}$ contains a countably infinite sequence of orthogonal elements.*
- Unitary orthoautomorphism: *For any two orthogonal atoms A and B there is an automorphism $\mathcal{U}$ such that $\mathcal{U}(A) = B$, which satisfies $\mathcal{U}(\overline{A}) = \overline{\mathcal{U}(A)}$, i.e., it is an* orthoautomorphism, *and whose mapping into $\mathcal{H}$ is a unitary operator U and therefore we also call it* unitary.

Thus we do arrive at a full Hilbert space, but the axioms for the Hilbert lattices that we used for this purpose are too involved to reveal a possible transition to its finite-dimensional representation. This is because in the past, the axioms were simply read off from the Hilbert space structure and were formulated as predicative statements of the first and second order that cannot be implemented by a quantum Turing machine. Besides, in this approach, quantum states can be defined only on the Hilbert space that we obtain at the end. Let us, instead, define them beforehand directly on an orthomodular lattice, so as to reduce to the Hilbert states when we add the axioms from Def. 3.11 to the lattice.

DEFINITION 3.14 *A state on a lattice $\mathcal{L}$ is a function $m : \mathcal{L} \longrightarrow [0,1]$ such that $m(1) = 1$ and $A \perp B \Rightarrow m(A \cup B) = m(A) + m(B)$, where $A \perp B$ means $A \leq \overline{B}$.*

This implies $m(A) + m(\overline{A}) = 1$ and $A \leq B \Rightarrow m(A) \leq m(B)$.

DEFINITION 3.15 *A nonempty set S of states on $\mathcal{L}$ is called a strong set of* classical *states if*

$$(\exists m \in S)(\forall A, B \in \mathcal{L})((m(A) = 1 \Rightarrow m(B) = 1) \Rightarrow A \leq B) \quad (3.126)$$

and a strong set of quantum *states if*

$$(\forall A, B \in \mathrm{L})(\exists m \in S)((m(A) = 1 \Rightarrow m(B) = 1) \Rightarrow A \leq B) \,. \quad (3.127)$$

We want to emphasize the difference between quantum and classical states. A classical state is the same for all lattice elements, while a quantum state might be different for each of the elements. The following theorem [Megill and Pavičić, 2000] shows us that a classical state can be be very strong.

THEOREM 3.16 *Any ortholattice that admits a strong set of classical states is distributive.*

Radoslaw Godowski [Godowski, 1981] found an infinite series of equations partly corresponding to the strong set of quantum states given by Eq. (3.127), forming a series of algebras contained in the class of all orthomodular lattices and containing the class of all Hilbert lattices. Megill and Pavičić [Megill and Pavičić, 2000] found another independent infinite series of equations partly corresponding to the remaining Hilbert algebraic structure and also forming a series of algebras contained in the class of all orthomodular lattices and containing the class of all Hilbert lattices. Such series can be cut at a chosen level and implemented by a Turing machine since they are of a zeroth (propositional) order. Whether it will be possible to write down Schrödinger equations by means of such series awaits future developments, as is the case with the majority of projects in quantum computing.

References

Abrams, Daniel S. and Lloyd, Seth (1997). Simulation of many-body fermi systems on a universal quantum computer. *Phys. Rev. Lett.*, **79**:2586–2589.

Abrams, Daniel S. and Lloyd, Seth (1999). Quantum algorithm providing exponential speed increase for finding eigenvalues and eigenvectors. *Phys. Rev. Lett.*, **83**:5162–5165.

Anonymous (1997). The millennium-bug muddle. *The Economist*, **345**(8037):17–18.

Azuma, Hiroo (2003). Interaction-free generation of entanglement. *Phys. Rev. A*, **68**:022320–1–10.

Azuma, Hiroo (2004). Interaction-free quantum computation. *Phys. Rev. A*, **70**: 012318–1–5.

Barrett, M. D., Chiaverini, J., Schaetz, T., Britton, J., Itano, W. M., Jost, J. D., Knill, E., Langer, C., Leibfried, D., Ozeri, R., and Wineland, David J. (2004). Deterministic quantum teleportation of atomic qubits. *Nature*, **429**:737–739.

Baxter, Rob and Trew, Arthur (2002). eDIKT: Harvesting knowledge through extreme IT. *Univ. Edinburgh EPCC News*, **46** (Summer):2.

Beenakker, C. W. J., DiVincenzo, D. P., Emary, C., and Kindermann, M. (2004). Charge detection enables free-electron quantum computation. *Phys. Rev. Lett.*, **93**:020501–1–4.

Beltrametti, Enrico G. and Cassinelli, Gianni (1981). *The Logic of Quantum Mechanics*. Addison-Wesley.

Bennett, Charles, Bethune, Donald, Brassard, Gilles, Donnangelo, Nicholas, Ekert, Artur, Elliott, Chip, Franson, James, Fuchs, Christopher, Goodman, Matthew, Hughes, Richard, Kwiat, Paul, Migdall, Alan, Nam, Sae-Woo, Nordholt, Jane, Preskill, John, and Rarity, John (2004). A quantum information science and technology roadmap. Part 2: Quantum computation; Report of the quantum cryptography technology experts panel. Technical Report LA-UR-04-4085, Los Alamos National Laboratory, Los Alamos.

Bennett, Charles H. (1973). Logical reversibility of computation. *IBM J. Res. Develop.*, **17**:525–532.

Bennett, Charles H. (1989). Time/space trade-offs for reversible computation. *S.I.A.M. J. Comp.*, **18**:766–776.

Bennett, Charles H. (1992). Quantum cryptography using any two nonorthogonal states. *Phys. Rev. Lett.*, **68**:3121–3124.

Bennett, Charles H. and Brassard, Gilles (1984). Quantum cryptography, public key distribution and coin tossing. In *International Conference on Computers, Systems & Signal Processing, Bangalore, India, December 10-12, 1984*, pages 175–179. IEEE, New York.

Bennett, Charles H., Brassard, Gilles, Popescu, Sandu, Schumacher, Benjamin, Smolin, John A., and Wootters, William K. (1996). Purification of noisy entanglement and faithful teleportation via noisy channels. *Phys. Rev. Lett.*, **76**:722–725.

Benson, Oliver, Santori, Charles, Pelton, Matthew, and Yamamoto, Yoshihisa (2000). Regulated and entangled photons from a single quantum dot. *Phys. Rev. Lett.*, **84**:2513–2516.

Benyoucef, M., Ulrich, S. M., Michler, P., Wiersig, J., Jahnke, F., and Forchel, A. (2004). Enhanced correlated photon pair emission from a pillar microcavity. *New J. Phys.*, **6**:91–1–12.

Beran, Ladislav (1985). *Orthomodular Lattices; Algebraic Approach.* D. Reidel, Dordrecht.

Bergmann, K., Theuer, H., and Shore, B. W. (1998). Coherent population transfer among quantum states of atoms and molecules. *Rev. Mod. Phys.*, **70**:1003–1025.

Bernstein, Ethan and Vazirani, Umesh (1997). Quantum complexity theory. *SIAM J. Comput.*, **26**:1411–1473.

Biham, Eli, Boyer, Michel, Boykin, P. Oscar, Mor, Tal, and Roychowdhury, Vwani (2000). A proof of the security of quantum key distribution. In *Proceedings of the Thirty-Second Annual ACM Symposium on Theory of Computing*, pages 715–724. ACM Press, New York.

Birkhoff, Garret and von Neumann, J. (1936). The logic of quantum mechanics. *Ann. Math.*, **37**:823–843.

Birnbaum, Joel and Williams, R. Stanly (2000). Physics and the information revolution. *Phys. Today*, **53** (1, January):38–42.

Blatt, R., Ertmer, W., Hall, J., and Zoller, P. (1986). Laser cooling of free atoms, a simulation approach. *Phys. Rev. A*, **34**:3022.

Boghosian, Bruce B. and Taylor IV, Washington (1998). Simulating quantum mechanics on a quantum computer. *Physica D*, **120**:30–42.

Born, Max and Wolf, Emil (1997). *Principles of Optics.* Cambridge University Press, Cambridge, 6th edition.

Bouwmeester, Dik, Pan, Jian-Wei, Mattle, Klaus, Eibl, Manfred, Weinfurter, Harald, and Zeilinger, Anton (1997). Experimental quantum teleportation. *Nature*, **390**:575–579.

Branning, David, Kwiat, Paul, and Migdall, Alan (2003). Obseervation of photons emitted during inhibited spontaneous emission. In Shapiro, Jeffrey H. and Hirota, Osamu, editors, *Proceedings of the Sixth International Conference on Quantum Communication, Measurement and Computing*, pages 129–132. Rinton Press, Princeton.

Braunstein, Samuel L., Caves, C. M., Jozsa, Richard, Linden, N., Popescu, Sandu, and Schack, R. (1999). Separability of very noisy mixed states and implications for NMR quantum computing. *Phys. Rev. Lett.*, **83**:1054–1057.

Braunstein, Samuel L. and Pati, Arun K., editors (2003). *Quantum Information with Continuous Variables.* Springer, Berlin–New York.

Brenner, Alfred E. (2001). More on Moore's Law. *Phys. Today*, **54**(7, July):84.

Bruss, Dagmar (1998). Optimal eavesdropping in quantum cryptography with six states. *Phys. Rev. Lett.*, **81**:3018–3021.

Buck, Joseph R. (2003). *Cavity QED in Microsphere and Fabry–Perot Cavities*. Ph.D. thesis. California Institute of Technology. Division of Engineering and Applied Science, Pasadena, CA.

Byrd, Mark (1998). Differential geometry on SU(3) with applications to three state systems. *J. Math. Phys.*, **39**:6125–6135.

Cabello, Adán (2000). Kochen–Specker theorem and experimental tests on hidden variables. *Int. J. Mod. Phys. A*, **15**:2813–2820.

Cabello, Adán, Estebaranz, José M., and García-Alcaine, Guillermo (1996). Bell–Kochen–Specker theorem: A proof with 18 vectors. *Phys. Lett. A*, **212**:183–187.

Cabello, Adán and García-Alcaine, Guillermo (1998). Proposed experimental tests of the Bell–Kochen–Specker theorem. *Phys. Rev. Lett.*, **80**:1797–1799.

Calderbank, A. R. and Shor, Peter W. (1996). Good quantum error-correcting codes exist. *Phys. Rev. A*, **54**:1098–1105.

Campos, Richard A., Saleh, Bahaa E. A., and Teich, Malvin C. (1989). Quantum-mechanical lossless beam splitter: SU(2) symmetry and photon statistics. *Phys. Rev. A*, **40**:1371–1384.

Cavallar, Stefania, Lioen, Walter M., te Riele, Herman J. J., Dodson, Bruce, Lenstra, Arjen K., Montgomery, Peter L., Murphy, Brian, Murphy, Brian, Aardal, Karen, Gilchrist, Jeff, Guillerm, Gérard, Leyland, Paul, Marchand, Joël, Morain, Françis, Muffett, Alec, Putnam, Chris, and Putnam, Craig (2000). Factorization of a 512-bit RSA modulus. In Preneel, Bart, editor, *Advances in Cryptology—EUROCRYPT 2000, International Conference on the Theory and Application of Cryptographic Techniques, Bruges, Belgium, May 14-18, 2000, Proceedings*, volume 1807 of *Lecture Notes in Computer Science*, pages 1–18. Springer, Berlin–New York.

Church, Alonzo (1936a). A note on the Entscheidungsproblem. *J. Symb. Logic*, **2**:42.

Church, Alonzo (1936b). A note on the Entscheidungsproblem: A correction. *J. Symb. Logic*, **2**:101–102.

Church, Alonzo (1936c). An unsolvable problem of elementary number theory. *Am. J. Math.*, **58**:345–363.

Cinelli, C., Barbieri, M., De Martini, F., and Mataloni, P. (2005). Realization of hyperentangled two-photon states. *Laser Physics*, **15**:124–128.

Cirac, J. Ignacio, Blatt, R., Parkins, A. S., and Zoller, Peter (1993). Preparation of Fock states of observation of quantum jumps in an ion trap. *Phys. Rev. Lett.*, **70**:762–765.

Cirac, J. Ignacio, Blatt, R., Zoller, Peter, and Phillips, W. D. (1992). Laser cooling of trapped ions in a standing wave. *Phys. Rev. A*, **46**:2668.

Cirac, J. Ignacio and Zoller, Peter (1995). Quantum computations with cold trapped ions. *Phys. Rev. Lett.*, **74**:4091–4094.

Cirac, J. Ignacio and Zoller, Peter (2000). A scalable quantum computer with ions in an array of microtraps. *Nature*, **404**:579–581.

Cirac, J. Ignacio and Zoller, Peter (2004). New frontiers in quantum information with atoms and ions. *Phys. Today*, **57**(3):38–44.

Clark, R. G., Brenner, R., Buehler, T. M., Chan, V., Curson, N. J., Dzurak, S., Gauja, E., Goan, H. S., Greentree, D., Hallam, T., Hamilton, R., Hollenberg, L. C. L., Jamieson, D. N., McCallum, J. C., Milburn, G. J., O'Brien, J. L., Oberbeck, L., Pakes, C. I., Prawer, S. D., Reilly, D. J., Ruess, F. J., Schofield, S. R., Simmons, M. Y., Stanley, F. E., Starrett, R. P., Wellard, C., and Yang, C. (2003). Progress in silicon-based quantum computing. *Phil. Trans. R. Soc. London A*, **361**:1451–1471.

Cleve, Richard, Ekert, Arthur, Macchiavello, Chiara, and Mosca, Michaele (1998). Quantum algorithms revisited. *Proc. R. Soc. Lond. A*, **454**:339–354.

Corndorf, Eric, Kanter, Gregory S., Liang, Chuang, and Kumar, Prem (2004). Data encryption over an inline-amplified 200-km-long WDM line using coherent-state quantum cryptography. In Donkor, Eric, Pirich, Andrew R., and Brandt, Howard E., editors, *Quantum Information and Computation II*, volume **5436** of *Proceedings of SPIE*, pages 12–20. SPIE, Bellingham, Washington.

Dalla Chiara, Maria Luisa (1986). Quantum logic. In Gabbay, D. and Guenthner, F., editors, *Handbook of Philosophical Logic*, volume **III.**, pages 427–469. D. Reidel, Dordrecht.

De Vos, Alexis, Desoete, B., Janiak, F., and Nogawski, A. (2001). Control gates as building blocks for reversible computers. In Pirsch, Peter and Hochet, Bertrand, editors, *Proceedings of the 11th International Workshop on Power and Timing Modeling, Optimization and Simulation, 2001*, pages 9201–9210. Yverdon.

De Vos, Alexis, Raa, Birger, and Storme, Leo (2002). Generating the group of reversible logic gates. *J. Phys. A*, **35**:7063–7078.

Degiorgio, Vittorio (1980). Phase shift between the transmitted and the reflected optical fields of a semireflecting lossless mirror is $\pi/2$. *Am. J. Phys.*, **48**:81–82.

Demtröder, Wolfgang (1996). *Laser Spectroscopy: Basic Concepts and Instumentation.* Springer, Berlin, 2nd edition.

Deutsch, David (1985). Quantum theory, the Church-Turing principle and the universal quantum computer. *Proc. R. Soc. Lond. A*, **400**:97–117.

Deutsch, David and Jozsa, Richard (1992). Rapid solutions of problems by quantum computation. *Proc. R. Soc. Lond. A*, **439**:553–558.

Devoret, Michel E., Esteve, Daniel, and Urbina, Christian (1992). Single-electron trensfer in metallic nanostructures. *Nature*, **360**(10 December):547–553.

Dishkant, Hermann (1974). The first order predicate calculus based on the logic of quantum mechanics. *Rep. Math. Logic*, **3**:9–18.

DiVincenzo, David P. (2000). The physical implementation of quantum computation. *Fortschr. Phys.*, **48**:771–783.

Dorai, Kavita, Arvind, and Kumar, Anil (2000). Implementing quantum-logic operations, pseudopure states, and the Deutsch-Jozsa algorithm using noncommuting selective pulses in NMR. *Phys. Rev. A*, 61:042306–1–70.

Ekert, Artur and Jozsa, Richard (1996). Quantum computation and Shor's factoring algorithm,. *Rev. Mod. Phys.*, **68**:733–753.

Elitzur, A. C. and Vaidman, Lev (1993). Quantum mechanical interaction-free measurements. *Found. Phys.*, **23**:987–997.

Elliott, C., Colvin, A., Pearson, D., Pikalo, O., Schlafer, J., and Yeh, H. (2005). Current status of the DARPA Quantum Network. In Donkor, E. J., Pirich, A. R., and Brandt, H. E., editors, *SPIE Quantum Information and Computation III*, volume **5815** of *Proceedings of SPIE*, pages 138–149. SPIE, Bellingham, Washington.

Fearn, H., Cook, R.J., and Milonni, P. W. (1995). Sudden replacement of a mirror by a detector in cavity QED: Are photons counted immediately? *Phys. Rev. Lett.*, **74**:1327–1330.

Feynman, Richard P. (1985). Quantum mechanical computers. *Optics News*, **11**(Feb.): 11–20.

Feynman, Richard P., Leighton, Robert B., and Sands, Mathew (1965). *The Feynman Lectures on Physics; Volume III. Quantum Mechanics.* Addison-Wesley, Reading, Massachusetts.

Fraenkel, Abraham A., Bar-Hillel, Yehoshua, and Levy, Azriel (1973). *Foundations of Set Theory.* North-Holland, Amsterdam.

Franson, James D. (1989). Noncontextual hidden variables and physical measurements. *Phys. Rev. Lett.*, **62**:2205–2208.

Gaubatz, U., Rudecki, P., Schiemann, S., and Bergmann, K. (1990). Population transfer between molecular vibrational levels by stimulated Raman scattering with partially overlapping laserfields. a new concept and experimental results. *J. Chem. Phys.*, **92**:5363–5376.

Gilchrist, A., White, A. G., and Munro, W. J. (2002). Entanglement creation using quantum interrogation. *Phys. Rev. A*, **66**:012106–1–7.

Gisin, Nicolas, Ribordy, Grégoire, Tittel, Wolfgang, and Zbinden, Hugo (2002). Quantum cryptography. *Rev. Mod. Phys.*, **74**:145–195.

Gödel, Kurt (1931). Über formal unentscheidbare Satze der Principia Mathematica und verwandter Systeme. *Monatash. Math. Physik*, **38**:173–198.

Gödel, Kurt (1934). *On Undecidable Propositions of Formal Mathematical Systems.* Lecture notes by Steven Cole Kleene and John Barkley Rosser. Institute for Advanced Study, Princeton, New Jersey.

Godowski, Radoslaw (1981). Varieties of orthomodular lattices with a strongly full set of states. *Demonstratio Math.*, **14**:725–733.

Gramss, Tino (1998). Solving the schrödinger equation for the feynman quantum computer. *Int. J. Theor. Phys.*, **37**:1423–1439.

Grangier, Philippe, Sanders, Barry, and Vuckovic, Jelena (Eds.) (2004). Focus on single photons on demand. *New J. Phys.*, **6**:85–100 (Editorial & 15 papers by over 70 authors).

Greiner, Walter (1989). *Quantum Mechanics*, volume **1**. An Introduction. Springer-Verlag, Berlin.

Gruska, Jozef (1997). *Foundations of Computing.* International Thomson Computer Press, Boston.

Gruska, Jozef (1999). *Quantum Computing.* McGraw-Hill, Boston.

Gulde, Stephan, Riebe, Mark, Lancaster, Gavin P. T., Becher, Christoph, Eschner, Jurgen, Haffner, Hartmut, Schmidt-Kaler, Ferdinand, Chuang, Isaac L., and Blatt, Rainer (2003). A silicon-based nuclear spin quantum computer. *Nature*, **421**:48–50.

Hammermesh, Morton (1962). *Group Theory and Its Application to Physical Problems.* Addison-Wesley, Reading, Mass.

Helms, Harry L., editor (1983). *The McGraw-Hill Computer Handbook.* McGraw-Hill, New York.

Herbrand, Jacques (1931). Sur le problème fondamental de la logique mathématique. *C. R. Varsovie*, **24**:12–56.

Hermes, Hans (1969). *Enumerability, Decidability, Computability.* Springer-Verlag, New York.

Hilbert, David and Ackermann, Wilhelm (1950). *Principles of Mathematical Logic.* Chelsea, New York.

Hilbert, David and Bernays, Paul (1934). *Grundlagen der Mathematik.* **I**. Springer, Berlin.

Hilbert, David and Bernays, Paul (1939). *Grundlagen der Mathematik.* **II**. Springer, Berlin.

Hioe, F. T. (1984). Linear and nonlinear constants of motion for two-photon processes in three-level sustems. *Phys. Rev. A*, **29**:3434–3436.

Holland, JR., Samuel S. (1995). Orthomodularity in infinite dimensions; a theorem of M. Solèr. *Bull. Am. Math. Soc.*, **32**:205–234.

Huang, J.-S. (1999). *Lectures on Representation Theory.* World Scientific, Singapore.

Huang, Yun-Feng, Li, Chuan-Feng, Yong-Sheng Zhang, Jian-Wei Pan, and Guo, Guang-Can (2002). Kochen–Specker theorem for finite precision spin-one measurement. *Phys. Rev. Lett.*, **88**:240402–1–4.

Hughes, R., Morgan, G., and Peterson, C. (2000). Quantum key distribution over a 48-km optical fiber network. *J. Mod. Opt.*, **47**:533–547.

Hughes, Richard, Doolen, Gary, Awschalom, David, Carlton, Chapman, Michael, Clark, Robert, Cory, David, DiVincenzo, David, Watson, Thomas J., Ekert, Artur, Hammel, P. Chris, Kwiat, Paul, Lloyd, Seth, Orlando, Gerard Milburn Terry, Steel, Duncan, Vazirani, Umesh, Whaley, K. Birgitta, and Wineland, David (2004). A quantum information science and technology roadmap. Part 1: Quantum computation; Report of the quantum information science and technology experts panel. Technical Report LA-UR-04-1778, Los Alamos National Laboratory, Los Alamos.

Hurst, S. (1984). Multiple-valued logic—its status and its future. *IEEE Trans. Comput.*, **33**:1160–1179.

Hwang, Won-Young (2003). Quantum key distribution with high loss: Toward global secure communication. *Phys. Rev. Lett.*, **91**:057901–1–4.

Jackson, John David (1967). *Classical Electrodynamics*. John Wiley & Sons, New-York.

James, D. F. V. (1998). Quantum dynamics of cold trapped ions with application to quantum computation. *Appl. Phys. B*, **66**:181–190.

Jozsa, Richard (2000). Quantum algorithms. In Bouwmeester, Dik, Ekert, Artur, and Zeilinger, Anton, editors, *The Physics of Quantum Information: Quantum Cryptography, Quantum Teleportation, Quantum Computation*, pages 104–126. Springer, Berlin.

Kalmbach, Gudrun (1974). Orthomodular logic. *Z. math. Logik Grundl. Math.*, **20**:395–406.

Kalmbach, Gudrun (1986). *Measures and Hilbert Lattices*. World Scientific, Singapore.

Kane, Bruce E. (1998). A silicon-based nuclear spin quantum computer. *Nature*, **393**:133–137.

Kane, Bruce E. (2000). Silicon-based quantum computation. *Fortschr. Phys.*, **48**:1023–1041.

Keller, Matthias, Lange, Birgit, Hayasaka, Kazuhiro, Lange, Wolfgang, and Walther, Herbert (2004a). A calcium ion in a cavity as a controlled single-photon source. *New J. Phys.*, **6**:95–1–22.

Keller, Matthias, Lange, Birgit, Hayasaka, Kazuhiro, Lange, Wolfgang, and Walther, Herbert (2004b). Continuous generation of single photons with controlled waveform in an ion-trap cavity system. *Nature*, **431**:1075–1078.

Kernaghan, Michael (1994). Bell–Kochen–Specker theorem for 20 vectors. *J. Phys. A*, **27**:L829–L830.

Kielpinski, D., Monroe, C., and Wineland, David J. (2002). Architecture for a large-scale ion-trap quantum computer. *Nature*, **417**:709–711.

Kim, Jungsang, Takeuchi, Shigeki, and Yamamoto, Yoshihisa (1999). Multiphoton detection using visible light photon counter. *Appl. Phys. Lett.*, **74**:902–904.

Kleene, Stephen Cole (1936). λ-definability and recursiveness. *Duke Math. J.*, **2**:340–353.

Kochen, Simon and Specker, E. P. (1967). The problem of hidden variables in quantum mechanics. *J. Math. Mech.*, **17**:59–87.

Kok, Pieter, Lee, Hwang, and Dowling, Jonathan P. (2002). Single-photon quantum-nondemolition detectors constructed with linear optics and projective measurements. *Phys. Rev. A*, **66**:063814–1–9.

Kok, Pieter, Williams, Colin P., and Dowling, Jonathan P. (2003). Construction of a quantum repeater with linear optics. *Phys. Rev. A*, **68**:022301–1–5.

Krazit, Tom (2004). Intel's pentium 4 to top out at 3.8 GHz. It's probably the last increase in this processor's clock speed. *PC World.* Online News, November 01.

Kuhn, Axel, Hennrich, Markus, Bondo, T., and Rempe, Gerhard (1999). Conrolled generation of single photons from a strongly coupled atom-cavity system. *App. Phys. B*, **69**:373–377.

Kuhn, Axel, Hennrich, Markus, and Rempe, Gerhard (2002). Deterministic single-photon source for distributed quantum networking. *Phys. Rev. Lett.*, **69**:067901–1–4.

Kuklinski, J. R., Gaubatz, U., Hioe, F. T., and Bergmann, K. (1989). Adiabatic population transfer in a three-level system driven by delayed laser pulses. *Phys. Rev. A*, **40**:6741–6744.

Kumar, Prem, Kwiat, Paul, Migdall, Alan, Nam, Sae Woo, Vuckovic, Jelena, and Wong, Franco N. C. (2004). Photonic technologies for quantum information processing. *Quantum Inform. Processing*, **3**:215–231.

Kwiat, P. G., White, A. G., Mitchell, J. R., Nairz, O., Weihs, G., Weinfurter, H., and Zeilinger, A. (1999). High-efficiency quantum interrogation measurements via the quantum Zeno effect. *Phys. Rev. Lett.*, **83**:4725–4728.

Kwiat, Paul G. (1997). Hyper-entangled states. *J. Mod. Opt.*, **44**:2173–2184.

Kwiat, Paul G., Mattle, Klaus, Weinfurter, Harald, Zeilinger, Anton, Sergienko, Alexander V., and Shih, Yanhua (1995). New high-intensity source of polarization-entangled photon pairs. *Phys. Rev. Lett.*, **75**:4337–4341.

Kwiat, Paul G. and Weinfurter, Harald (1998). Embedded Bell-state analysis. *Phys. Rev. A*, **58**:R2623–R2626.

Laflamme, Raymond, Knill, Emanuel, Cory, David G., Fortunato, Evan M., Havel, Timothy F., Miquel, Cesar, Martinez, Rudy, Negrevergne, Camille J., Ortiz, Gerardo, Pravia, Marco A., Sharf, Yehuda, Sinha, Suddhasattwa, Somma, Rolando, and Viola, Lorenza (2002). NMR and quantum information processing. *Los Alamos Science*, (**27**):226–259.

Langholz, Gideon, Francioni, Joan, and Kandel, Abraham (1989). *Elements of Computer Organization.* Prentice-Hall, London.

Levy, Jeremy (2002). Universal quantum computation with spin-1/2 pairs and Heisenberg exchange. *Phys. Rev. Lett.*, **89**:147902–1–3.

Lloyd, Seth (1996). Universal quantum simulators. *Science*, **273**:1073–1078. Correction, *Ibid.* **279**, 1117 (1998).

Lloyd, Seth and Braunstein, Samuel L. (1999). Quantum computation over continuous variables. *Phys. Rev. Lett.*, **82**:1784–1787.

Lo, Hoi-Kwong and Chau, H. F. (1999). Unconditional security of quantum key distribution over arbitrarily long distances. *Science*, **283**:2050–2056.

Lo, Hoi-Kwong, Chau, H. F., and Ardehali, M. (2005). Efficient quantum key distribution scheme and a proof of its unconditional security. *J. Cryptology*, **82**:133–166.

Löfler, Markus, Meyer, Georg M., and Walther, Herbert (1997). Spectral properties of one-atom laser. *Phys. Rev. A*, **55**:3923–3930.

Lütkenhaus, N., Calsamiglia, J., and Suominen, K.-A. (1999). Bell measurements for teleportation. *Phys. Rev. A*, **59**:3295–3300.

Mackey, George Whitelaw (1963). *The Mathematical Foundations of Quantum Mechanics.* W. A. Benjamin, New York.

MacLaren, M. Donald (1964). Atomic orthocomplemented lattices. *Pacif. J. Math.*, **14**:597–612.

MacWilliams, F. J. and Sloane, N. J. A. (1977). *The Theory of Error-Correcting Codes.* North-Holland, Amsterdam.

Maeda, Fumitomo and Maeda, Shûichirô (1970). *Theory of Symmetric Lattices.* Springer-Verlag, New York.

Markov, Andrei Andreevich (1947). The impossibility of certain algorithms in the theory of associative systems. *Soviet Mathematics—Doklady,* **55**:583–586.

Markov, Andrei Andreevich (1961). *Theory of Algorithms.* Keter Press, Jerusalem. Originally published in Russian in Works of the Mathematical Institute V. A. Steklov, Vol. XLII, Academy of Sciences of the USSR, Moskva-Leningrad, 1954.

Mayers, Dominic (2001). Unconditional security in quantum cryptography. *JACM,* **48**:351–406.

McCune, William, Veroff, R., Fitelson, B., Harris, K., Feist, A., and Wos, L. (2002). Short single axioms for Boolean algebra. *J. Automated Reasoning,* **29**:1–16.

McKay, Brendan D. (1998). Isomorph-free exhaustive generation. *J. Algorithms,* **26**:306–324.

McKay, Brendan D., Megill, Norman D., and Pavičić, Mladen (2000). Algorithms for Greechie diagrams. *Int. J. Theor. Phys.,* **39**:2381–2406.

McKeever, J., Boca, A., Boozer, A. D., Miller, R., Buck, J. R., Kuzmich, A., and Kimble, H. J. (2004). Deterministic generation of single photons from one atom trapped in a cavity. *Science,* **303**:1992–1994.

Megill, Norman D. and Pavičić, Mladen (2000). Equations, states, and lattices of infinite-dimensional Hilbert space. *Int. J. Theor. Phys.,* **39**:2337–2379.

Messiah, Albert (1965). *Quantum Mechanics.* North-Holland, Amsterdam.

Mitchison, Graeme and Jozsa, Richard (2001). Counterfactual computation. *Proc. R. Soc. Lond. A,* **457**:1175–1193.

Mittelstaedt, Peter (1978). *Quantum Logic.* Synthese Library; Vol. 18. Reidel, London.

Monroe, C., Leibfried, D., King, B. E., Meekhof, D. M., Itano, W. M., and Wineland, D. J. (1997). Simplified quantum logic with trapped ions. *Phys. Rev. A,* **55**:R2489–R2491.

Mu, Yi and Savage, C. M. (1992). One-atom laser. *Phys. Rev. A,* **46**:5944–5954.

Nielsen, Michael A. and Chuang, Isaac L. (2000). *Quantum Computation and Quantum Information.* Cambridge University Press, Cambridge.

Ou, Z. Y. and Mandel, L. (1989). Derivation of reciprocity relations for a beam splitter from energy balance. *Am. J. Phys.,* **57**:66–67.

Ouellette, Jannifer (2005). Quantum key distribution. *Industr. Physicist,* (Dec 2004/ Jan 2005):22–25.

Padgett, Miles, Courtial, Johannes, and Allen, Les (2004). Light's orbital angular momentum. *Phys. Today,* **57**(5):35–40.

Pan, Jian-Wei, Bouwmeester, Dik, Weinfurter, Harald, and Zeilinger, Anton (1998). Experimental entanglement swapping: Entangling photons that never interacted. *Phys. Rev. Lett.,* **80**:3891–4894.

Pan, Jian-Wei, Simon, Christoph, Brukner, Časlav, and Zeilinger, Anton (2001). Experimental purification for quantum communication. *Nature,* **410**:1067–1070.

Pan, Jian-Wei and Zeilinger, Anton (1998). Greenberger-Horne-Zeilinger-state analyzer. *Phys. Rev. A,* **57**:2208–2211.

Paul, Harry and Pavičić, Mladen (1996). Resonance interaction-free measurement. *Int. J. Theor. Phys.,* **35**:2085–2091.

Paul, Harry and Pavičić, Mladen (1997). Nonclassical interaction-free detection of objects in a monolithic total-internal-reflection resonator. *J. Opt. Soc. Am. B,* **14**:1273–1277.

Paul, Harry and Pavičić, Mladen (1998). Realistic interaction-free detection of objects in a resonator. *Found. Phys.*, **28**:959–970.

Pavičić, Mladen (1986). *Algebraico-Logical Structure of the Interpretations of Quantum Mechanics.* Ph.D. thesis (in Croatian). University of Belgrade, Belgrade.

Pavičić, Mladen (1987a). Complex Gaussians and the Pauli non-uniqueness. *Phys. Lett. A*, **122**:280–282.

Pavičić, Mladen (1987b). Minimal quantum logic with merged implications. *Int. J. Theor. Phys.*, **26**:845–852.

Pavičić, Mladen (1989). Unified quantum logic. *Found. Phys.*, **19**:999–1016.

Pavičić, Mladen (1994). Spin correlated interferometry for polarized and unpolarized photons on a beam splitter. *Phys. Rev. A*, **50**:3486–3491.

Pavičić, Mladen (1995). Spin-correlated interferometry with beam splitters: Preselection of spin-correlated photons. *J. Opt. Soc. Am. B*, **12**:821–828.

Pavičić, Mladen (1996). Resonance energy-exchange-free detection and 'welcher Weg' experiment. *Phys. Lett. A*, **223**:241–245.

Pavičić, Mladen (1997). A method for reaching detection efficiencies necessary for optical loophole-free Bell experiments. *Optics Commun.*, **142**:308–314.

Pavičić, Mladen (2000a). Quantum simulators and quantum repeaters. *Fortschr. Phys.*, **48**:497–503.

Pavičić, Mladen (2000b). A realistic interaction-free resonator. In Kumar, Prem, D'Ariano, G. M., and Hirota, Osamu, editors, *Quantum Communication, Computing, and Measurement 2*, pages 527–531. Kluwer/Plenum, New York.

Pavičić, Mladen (2002). Quantum computers, discrete space, and entanglement. In Callaos, Nagib, He, Yigaang, and Perez-Peraza, Jorge A., editors, *SCI 2002/ISAS 2002 Proceedings, The 6th World Multiconference on Systemics, Cybernetics, and Informatics*, volume **XVII**, SCI in Physics, Astronomy and Chemistry, pages 65–70. SCI, Orlando, Florida.

Pavičić, Mladen and Megill, Norman D. (1999). Non-orthomodular models for both standard quantum logic and standard classical logic: Repercussions for quantum computers. *Helv. Phys. Acta*, **72**:189–210.

Pavičić, Mladen, Merlet, Jean-Pierre, McKay, Brendan D., and Megill, Norman D. (2005). Kochen–Specker vectors. *J. Phys. A*, **38**:497–503. Corrigendum, *J. Phys. A* **38**:3709 (2005).

Pavičić, Mladen and Summhammer, Johann (1994). Interferometry with two pairs of spin correlated photons. *Phys. Rev. Lett.*, **73**:3191–3194.

Pellizzari, Thomas (1997). Quantum networking with optical fibres. *Phys. Rev. Lett.*, **79**:5242–5245.

Pellizzari, Thomas (1998). Quantum computers, error-correction and networking: Quantum-optical approaches. In Kwong Lo, Hoi, Popescu, Sandu, and Spiller, Tim, editors, *Introduction to Quantum Computation and Information*, pages 270–310. World Scientific, Singapore.

Pittenger, Arthur O. (1999). *An Introduction to Quantum Computing Algorithms*, volume 19 of *Progress in Computer Science and Applied Logic.* Birkhäuser, Boston.

Pittman, T. B., Jacobs, B. C., and Franson, J. D. (2001). Probabilistic quantum logic operations using polarizing beam splitters. *Phys. Rev A*, **64**:062311–1–9.

Pomerance, Carl (1996). A tale of two sieves. *Notices of the AMS*, **43**:1473–14.

Post, Emil Leon (1936). Finite combinatory processes—formulation 1. *J. Symb. Logic*, **1**:103–105.

Post, Emil Leon (1943). Formal reduction of general combinatorial decision problem. *Am. J. Math.*, **65**:197–215.

Poyatos, J.F., Cirac, J.I., and Zoller, P. (2000). Schemes of quantum computations with trapped ions. *Fortschr. Phys.*, **48**:785–799.

Preskill, John (1998). Reliable quantum computers. *Proc. R. Soc. Lond. A*, **454**:385–410.

Pták, Pavel and Pulmannová, Sylvia (1991). *Orthomodular Structures as Quantum Logics.* Kluwer, Dordrecht.

Reinsch, Matthias W. (2000). A simple expression for the terms in the Baker-Campbell-Hausdorff series. *J. Math. Phys.*, **41**:2434–2442.

Rivest, Ronald L., Shamir, Adi, and Adleman, Leonard A. (1978). A method for obtaining digital signatures and public-key cryptosystems. *Commun. ACM*, **21**(2):120–126.

Saleh, Bahaa E. A. and Teich, Malvin Carl (1991). *Fundamentals of Photonics.* John Wiley & Sons, New York.

Sarkisyan, D., Papoyan, A., Varzhapetyan, T., Blushs, K., and Auzinsh, M. (2005). Fluorescence of rubidium in a submicrometer vapor cell: Spectral resolution of atomic transitions between Zeeman sublevels in a moderate magnetic field. *J. Opt. Soc. Am. B*, **22**:88–95.

Schenkel, T., Persaud, A., Park, S. J., Nilsson, J., Bokor, J., Liddle, J. A., Keller, R., Schneider, D. H., Cheng, D. W., and Humphries, D. E. (2003). Solid state quantum computer development in silicon with single ion implantation. *J. Appl. Phys.*, **94**:7017–7024.

Scully, M. O., Englert, B.-G., and Walther, Herbert (1991). Quantum optical test of complementarity. *Nature*, **351**:111–116.

Shor, P. W. (1997). Polynomial-time algorithms for prime factorization and discrete logarithms on a quantum computer. *SIAM J. Comp.*, **26**:1484–1509.

Shor, Peter (1994). Algorithms for quantum computation: Discrete logarithms and factoring. In Goldwasser, Shafi, editor, *Proceedings of the 35th Annual Sympsium on Foundations of Computer Science, November 20-22, 1994, Santa Fe, New Mexico*, pages 124–134, Los Alamitos, California. IEEE Computer Society Press.

Shor, Peter W. and Preskill, John (2000). Simple proof of security of the BB84 quantum key distribution protocol. *Phys. Rev. Lett.*, **85**:441–444.

Simon, Daniel R. (1997). On the power of quantum computation. *SIAM J. Comput.*, **26**:1474–1483.

Skinner, A. J., Davenport, M. E., and Kane, Bruce E. (2003). Hydrogenic spin quantum computing in silicon: a digital approach. *Phys. Rev. Lett.*, **90**:087901.

Solèr, Maria Pia (1995). Characterization of Hilbert spaces by orthomodular spaces. *Comm. Alg.*, **23**:219–243.

Sørensen, Anders and Mølmer, Klaus (1999). Quantum computation with ions in thermal motion. *Phys. Rev. Lett.*, **82**:1971–1974.

Sørensen, Anders and Mølmer, Klaus (2000). Entanglement and quantum computation with ions in thermal motion. *Phys. Rev. A*, **62**:022311–1–11.

Stace, T. M., Milburn, G. J., and Barnes, C. H. W. (2003). Entangled two-photon source using biexciton emission of an asymmetric quantum dot in a cavity. *Phys. Rev. B*, **67**:085317–1–15.

Steane, Andrew (1996a). Error correcting codes in quantum theory. *Phys. Rev. Lett.*, **77**:793–797.

Steane, Andrew (1996b). Multiple-particle interference and quantum error correction. *Proc. R. Soc. London A*, **452**:2551–2577.

Steane, Andrew (1997). The ion trap quantum information processor. *Appl. Phys. B*, **64**:623–642.

Steane, Andrew (1998). Quantum computing. *Rep. Prog. Phys.*, **61**:117–173.

Summhammer, Johann (1997). Factoring and Fourier transformation with a Mach-Zehnder interferometer. *Phys. Rev. A*, **56**:4324–4326.

Tapster, Paul R., Rarity, John G., and Owens, P. C. M. (1994). Violation of Bell's inequality over 4 km of optical fiber. *Phys. Rev. Lett.*, **73**:1923–1926.

Toffoli, Tommaso (1980). Reversible computing. In de Bakker, J. W. and van Leeuwen, Jan, editors, *Automata, Languages and Programming, 7th Colloquium, Noordweijkerhout, The Netherland, July 14-18, 1980, Proceedings*, volume **85** of *Lecture Notes in Computer Science*, pages 632–644. Springer, Berlin.

Tolman, Richard C. (1924). Duration of molecules in upper quantum states. *Phys. Rev.*, **23**:693–709.

Townsend, Paul, Rarity, John, and Tapster, Paul (1993). Single photon interference in a 10 km long optical fiber interferometer. *Electron. Lett.*, **29**:634–639.

Tsegaye, T., Goobar, E., Karlsson, Anders, Björk, Gunnar, Loh, M. Y., and Lim, K. H. (1998). Efficient interaction-free measurements in a high-finesse interferometer. *Phys. Rev. Lett.*, **57**:3987–3990.

Turing, Alan Mathison (1936). On computable numbers, with an application to the Entscheidungsproblem. *Proc. London Math. Soc., Ser. 2*, **42**:230–265.

Turing, Alan Mathison (1937). On computable numbers, with an application to the Entscheidungsproblem: A correction. *Proc. London Math. Soc., ser. 2*, **43**:544–546.

Turing, Alan Mathison (1948). "Intelligent machinery." national physical laboratory report. In Meltzer, B. and Michie, D., editors, *Machine Intelligence 5*. Edinburgh University Press, Edinburgh, 1969.

Turing, Alan Mathison (1950). Computing machinery and intelligence. *Mind*, **59**:433–460.

Vaidman, Lev and Yoran, Nadav (1999). Methods for reliable teleportation. *Phys. Rev. A*, **59**:116–125.

Vandersypen, Lieven M. K., Steffen, Matthias, Breyta, Gregory, Yannoni, Costantino S., Sherwood, Mark H., and Chuang, Isaac L. (2001). Experimental realization of Shor's quantum factoring algorithm using nuclear magnetic resonance. *Nature*, **414**:883–887.

Varadarajan, V. S. (1968,1970). *Geometry of Quantum Theory, Vols. 1 & 2.* John Wiley & Sons, New-York.

Weidman, Joachim (1980). *Linear Operators in Hilbert Spaces.* Springer-Verlag, New York.

Weigert, Stefan (1992). Pauli problem for a spin of arbitrary length: A simple method to determine its wave function. *Phys. Rev. A*, **45**:7688–7696.

Weisstein, Eric W. (2003). RSA-576 factored. *Wolfram MathWorld Headline News*, (December 5, 2003).

Wineland, David J., Barrett, M., Britton, J., Chiaverini, J., DeMarco, B., Itano, W. M., Jelenković, B., Langer, C., Leibfried, D., Meyer, V., Rosenband, T., and Schätz, T. (2003). Quantum information processing with trapped ions. *Phil. Trans. R. Soc. Lond. A*, **361**:1349–1361.

Wineland, David J., Dalibard, J., and Cohen-Tannoudji, C. (1992). Sisyphus cooling of a bound atom. *J. Opt. Soc. Am. B*, **9**:32–42.

Wineland, David J. and Itano, Wayne M. (1979). Laser cooling of atoms. *Phys. Rev. A*, **20**:1521–1540.

Wineland, David J., Monroe, C., Itano, W. M., Leibfried1, D., King, B. E., and Meekhof, D. M. (1998). Experimental issues in coherent quantum-state manipulation of trapped atomic ions. *J. Res. Natl. Inst. Stand. Techn.*, **103**:259–328.

Wolfram, Stephen (2002). *A New Kind of Science*. Wolfram Media, Champaign, IL.

Yang, Dori Jones (2000). On Moore's Law and fishing: Gordon Moore speaks out. *U. S. News Online*, (7 October).

Yu, Bo, Zhou, Zheng-Wei, Zhang, Yong, Xiang, Guo-Yong, and Guo, Guang-Can (2004). Robust high-fidelity teleportation of an atomic state through the detection of cavity decay. *Phys. Rev A*, **70**:014302–1–4.

Zalka, Christof (1998). Simulating quantum systems on a quantum computer. *Proc. Roy. Soc. London A*, **454**:313–322.

Zhao, Zhi, Yang, Tao, Chen, Yu-Ao, Zhang, An-Ning, and Pan, Jian-Wei (2003). Experimental realization of entanglement concentration and a quantum repeater. *Phys. Rev. Lett.*, **90**:207901–1–4.

Zimba, Jason and Penrose, Roger (1993). On Bell non-locality without probabilities: More curious geometry. *Stud. Hist. Phil. Sci.*, **24**:697–720.

Zwiller, Valery, Aichele, Thomas, and Benson, Oliver (2004). Quantum optics with single quantum dot devices. *New J. Phys.*, **6**:96.

Index

Pavičić, Mladen [From MARQUISE *Who's Who in Science and Engineering*, 4th Ed., 1998-1999 and *Who's Who in the World*, 17th, 2000] Physicist, educator; b. Zagreb, Croatia; PhD in Physics, 1986; main asst. Univ. Zagreb, 1982–89, asst. prof., 1990–95, assoc. prof., 1996–2000, full prof., 2001– ; head sci. project Ministry of Sci., Zagreb, 1991–96; head sci. project *Quantum Computation and Quantum Communication*, Ministry of Sci., Zagreb, 1996–2001, head sci. project *Quantum Information Theory*, Ministry of Sci., Educ., and Sport, Zagreb, 2001–. Author: *Solved Problems in Physics*, 1982, 2nd edit., 1984; articles to prof. journals: *Phys. Rev. Lett.; Phys. Rev. A, D; J. Opt. Soc. Am. B; Opt. Commun.; J. Phys. A; Phys. Lett. A; Helv. Phys. Acta; Forschr. Phys.; Found. Phys.; Int. J. Theor. Phys.;* etc.; Grantee Alexander von Humboldt Found., Germany, Univ. Cologne, Germany, 1988–90; Tech. Univ. Berlin, Germany, 1993; Humboldt Univ. Berlin, Germany, 1995; Grantee Fulbright Senior Scholar Research/Lect., USA, Univ. Maryland Baltimore County, UMBC, Baltimore, USA, 1999–2000; French Ministry Sci., Univ. Reims, France, 1992; Austrian Ministry Sci., Atom-Inst. of Austr. Univ., Vienna, Austria, 1993, 94, 95, 97; Max-Planck Gesel., Germany, Humboldt Univ. Berlin, Germany, 1996; E. Schrödinger Inst., Vienna, Austria, 2000; Mem. Int. Quantum Structures Assn. (co-founder, nominating com. 1992–94), USA; Croatian Humboldt-Club (president, 2002–); Croatian Phys. Soc.; European Phys. Soc.; Am. Phys. Soc.; Opt. Soc. Am.; Achievements include proof of Pauli nonuniqueness for real states, discovery of polarization correlation between beams of unpolarized light, discovery of polarization entanglement of two unpolarized photons that nowhere interacted, discovery of a nondistributive model for classical logic, co-discovery of a nonorthomodular model for quantum logic, co-formulation of a resonator interaction-free detection, discovery of an interaction free destruction of atom interference pattern, and co-discovery of exhaustive algorithms for generating arbitrary Kochen–Specker vectors.